EXPERIMENTAL CILIATOLOGY

EXPERIMENTAL CILIATOLOGY

An Introduction to Genetic and Developmental Analysis in Ciliates

D. L. NANNEY

Department of Genetics and Development
University of Illinois, Urbana

A WILEY-INTERSCIENCE
PUBLICATION

JOHN WILEY & SONS

New York Chichester
Brisbane Toronto

Library of Congress Cataloging in Publication Data

Nanney, David Ledbetter, 1925–
   Experimental ciliatology.

   "A Wiley-Interscience publication."
   Includes index.
   1. Ciliata—Genetics.   2. Ciliata—Development.   3. Genetics—Animal models.   4. Protozoa—
Genetics.   5. Protozoa—Development.   I. Title.

QL368.A22N36   593´.172´0415   79-21918
ISBN 0-471-06008-9

Printed in the United States of America

10 9 8 7 6 5 4 3 2 1

For PEARL and GRADY

and RUTH and TRACY

and (most of all)

for JEAN

Foreword

This is a unique book, written by an outstandingly qualified author. The book shows how the life-style and qualities of an organism set its uses as a tool for experimentation. Conclusions are reached after comparing, whenever possible, results obtained on several species of a related group of organisms, the ciliates, whose life-styles are particularly suitable for experimental analyses of certain aspects of developmental and genetic biology. The book is a substantial introduction to what has been, is being, and should continue to be contributed to these currently lively subjects by studies on these organisms. The introduction assumes only the most elementary knowledge of biology, and no knowledge of ciliates. Nevertheless, the author is so thoughtful and original that even this introduction is full of insights and ideas that make it *must* reading for the expert. The combination of all of these features in one book is, to my knowledge, unique. To top it all, the book is well-written, thanks perhaps to the author's life-long interest in English literature.

Professor Nanney began his biological career, shortly after World War II, at Indiana University. It was then a leading center of classical, microbial, and ciliate genetics, attracting brilliant and ambitious students of genetics who interacted excitedly with each other and with the faculty. In this atmosphere, Nanney caught fire and his flame has burned brightly ever since. With two of the five organisms discussed in this book, he has worked intensively; on most of the topics dealt with in the book, he has published several discoveries and interpretations. His earliest studies were on the genetics of mating types and killers in the then most fully "domesticated" ciliate, Paramecium. Soon, however, he and his co-workers set out to domesticate and exploit Tetrahymena for genetic work. This effort resulted in an impressive mass of data and insights into many topics: mating types; immobilization antigens (serotypes); cell surface structure and geometry; clonal development and aging; inbreeding degeneration; composition of the species complex and its remarkable structural conservatism, along with great molecular divergence; and determination and organization of the distinctive somatic nucleus of ciliates. Tetrahymena thus became one of the most popular, useful, and important ciliates for certain kinds of developmental, genetic, and molecular experimentation. Some believe it is destined to become the *Escherichia coli* of eukaryotic cells. To whatever extent this comes to pass, it will be traceable in large measure to the basic discoveries, stimulating ideas, and inspiration of D. L. Nanney.

Although the book fully portrays the importance of the work on Tetrahymena, it is committed to the comparative point of view and justifies it by, among other things, recounting very important types of work on other ciliates that has not and probably could not be performed on Tetrahymena. Such broad, comparative knowledge of ciliate development and genetics, though rich in significance for general biology, has until now been unduly limited to specialists who have talked mainly to each other. The present book attempts—fascinatingly, it seems to me—to make available to all biologists the highlights of this knowledge and its significance for general biology.

T. M. SONNEBORN

Geneva, Switzerland
August 1979

Contents

ix

EXPERIMENTAL
CILIATOLOGY

Introduction 1

THE CILIATED PROTOZOA HAVE BEEN THE OBJECT OF experimental analysis for about a century ever since improvements in staining procedures and microscope construction made such studies feasible. This research has continued in a steady stream and has produced a remarkably rich literature. Yet this heritage of understanding has been frustratingly difficult to acquire. The frustration of nonspecialists who want to understand the material has been matched or exceeded by the frustration of ciliate investigators who want their work appreciated outside the narrow walls of their own discipline. The present work is a modest effort to alleviate those frustrations.

The purpose of this book is not, however, simply to gratify the need of protozoologists for appreciation, and the desire of curious nonprotozoologists for information. An equally substantial motive is to awaken a desire for understanding among the nonprotozoologists and to explain why a biologist who is not a student of the protozoa should make the effort to comprehend a group of organisms that sometimes challenges our assumptions and violates our expectations.

These motives explain much about this effort, but certain deepseated prejudices about teaching and research should also be exposed at the outset. The first of these prejudices concerns the value of the comparative approach in biology. We assume, because of their shared structures and functions, that all life forms have a common origin. Particularly, we are impressed by the use of precisely the same amino acids and the same purines and pyrimidines in the macromolecules of all organisms. Any student of elementary organic chemistry can, and usually does, invent a few new amino acids or a new kind of nucleoside. Yet organisms have generally resisted this temptation—over enormous stretches of geological time. While it is true that organic similarities can be explained on the basis of necessity or of convergence, instead of common descent, these explanations are increasingly difficult to accept as more and more details of molecular mechanisms are found to conform in circumstantial detail in most kinds of living things.

This common origin, as well as similarity of structure, permits us to study biological processes in the most convenient life form with considerable assurance that the understanding achieved with one organism will be applicable to many others. Confidence in this extrapolability of understanding increases with increasing evidence for the conservatism of life. The consistent use of the same 20 amino acids in the construction of proteins is only one piece of evidence for such conservatism. The codon dictionary and the general procedures for synthesizing proteins apparently have remained without essential change ever since protein synthesis was "invented" and perfected by some enterprising ancestor.

Another well-known example of such conservatism is the cilium and its homologous structures—the flagellum of algae and spermatozoa (but not of bacteria), the basal bodies, and the centrioles of the mitotic apparatus. The 9 + 2 pattern is stable, nearly as stable as the codon dictionary, and perhaps as arbitrary in some respects. Why should only 20 amino acids be encoded? Why this particular set of 20? Why should the cilium have 9 microtubular arrays in the outer ring instead of 7 or 10? Neither the protein factory nor the cilium may be a "perfect" solution to the particular organismic function it serves, and either may be somewhat arbitrary, but their stability cannot be denied. Apparently, once these basic biological inventions were perfected, they became isolated on an adaptive peak, unable to be substantially "improved" by the single step changes that mark the usual meandering and opportunistic path of evolution. The striking diversities among organisms do not reside primarily in their basic biological structures and functions, but rather in the patterns of integration of essentially common life elements.

The major constraint on the comparative approach concerns the time and place in evolutionary history at which particular inventions occurred. The cilium (or the chromosome) cannot be studied in a prokaryote. No bacterium or blue-green alga contains a true flagellum. The cilium is, however, distributed almost universally among the eukaryotes and must have appeared first in their common ancestor; and it has persisted without essential change in all the descendants of that ancestor. Where the cilium is studied is immaterial, whether in the alga Chlamydomonas or in the mammalian oviduct, but the cilium cannot be studied in organisms that have not descended from the primitive but inventive, eukaryote parent. For similar reasons, it may not be possible to discover and analyze in unicellular organisms the major inventions that permitted the evolution of the multicellular state, if indeed such inventions exist. Perhaps, however, multicellular development, mentality, and social organization are the results of diverse patterns of integration of the elementary eukaryote (and prokaryote) mechanisms. Perhaps no "new" mechanism is to be found—only a better understanding of the system of integration.

This question of special multicellular mechanisms may not be answered with assurance until we have completed an inventory of cellular mechanisms, and we have little basis for guessing how near to completion that task is. A modern fable tells of a boy looking at night for a lost coin beneath a streetlight, even though the coin had been lost elsewhere. His explanation is that he can see only beneath the streetlight. Much of modern biology is clustered around a relatively few light sources—organisms such as *Escherichia coli* and *Drosophila melanogaster,* techniques such as breeding analysis, electron microscopy, density gradients, and electrophoresis. Some coins will not be collected here, however. We need new techniques, new questions, new organisms—not primarily because unique processes will be discovered (though they may), but because common mechanisms are likely to be illuminated in new places.

A commitment to comparative biology and a belief in the significance of the ciliate experimental heritage explain why this book was written, but they do not explain how it was written.

Obviously, an introduction is not an encyclopedia; choices of subject matter are required and some justification of those choices is expected.

My first impulse was to reduce to a minimum the jargon of the discipline and to try to communicate in the biological *lingua franca*—the "standard" language of the text book and the classroom. Although I have some sympathy with this tactic, my experience caused me to resist. Recently a chemist colleague, in a debate over foreign language requirements for undergraduates, pointed out that a beginning student in chemistry is required in the first semester to master as large a new vocabulary as is a beginning student in Spanish or Greek. Peter Yankwich was arguing that foreign language study is not just a kind of academic discipline, but indeed the epitome of academic discipline. All areas of understanding must be entered by routes marked by new and specialized terms describing things, processes, and relationships. One has to develop a vocabulary and a grammar of structures, mechanisms, and phenomena before one can think about the subject matter of the discipline.

The language of "standard biology" does not equip students to enter the ciliate world directly. Indeed "standard biology" is an Esperanto device—an artificial language derived from the major biological dialects, but inadequate for effective communication in many areas. We simply cannot express ciliate phenomena without recourse to ciliate vocabulary. This vocabulary is not now a part of the undergraduate biological heritage. Most students' experiences of ciliates begin, and end, with an introduction to Paramecium in the first laboratory exercise. They may be supplemented with occasional vague and often inaccurate allusions to simplicity or immortality.

For this reason, one of the intentions of the present work is to introduce an elementary "language" of protozoan structures and processes, the essential fabrics and phenomena that compose the life of a unique group of organisms. This introduction makes no attempt to be complete, and the specialist must accept my apologies for somewhat arbitrary choices. My own inadequacies keep me away from some areas—ecology, nutrition, taxonomy, and ultrastructure. I am nevertheless convinced that to understand the ciliates, or indeed any organisms, one must see them whole, that is, from a number of different perspectives. We are entering an era of rising concern with putting organisms back together, under the expectation that genetics can illuminate biochemistry, with an understanding that development has evolved. I hope that a whole built of a limited array of inadequate parts may nevertheless seduce and provoke students of biology into a better understanding of an interesting and useful group of organisms, and through this understanding into a better comprehension of life itself.

SUMMARY

The ciliated protozoa claim the attention of the general biologist for several reasons. They manifest most of the fundamental eukaryotic "inventions," they imbed those inventions in a unique biological matrix, and experimental analyses of the inventions and their organization lead to conclusions applicable to organisms more difficult to examine.

To understand the studies on the ciliates one must see them as four-dimensional organisms, not as "generalized" eukaryotes. One must become acquainted with their structure, behavior, and evolutionary history before one is competent to separate the general and the particular, and to evaluate their contributions judiciously.

GENERAL WORKS CONCERNED WITH THE CILIATES

Calkins, G. N. and F. M. Summers (Ed.) 1941. *Protozoa in Biological Research,* Reprinted 1964, Hafner.

Chen, T. T. (Ed.) 1967–1972. *Research in Protozoology,* 4 vols., Pergamon.

Corliss, J. O. 1979. *The Ciliated Protozoa,* Second Edition, Pergamon.

Grell, K. G. 1973. *Protozoology,* Springer-Verlag.

Hutner, S. H. 1975. Maintaining protozoa and protozoan diversity. In *The Role of Culture Collections in the Era of Molecular Biology* (R. R. Colwell, Ed.) American Society of Microbiology, pp. 43–52.

Jahn, T. L., E. C., Bovee, and F. F. Jahn 1979. *How to Know the Protozoa,* Second Edition, Brown.

Jones, A. R. 1974. *The Ciliates,* St. Martins.

Kudo, R. R. 1966. *Protozoology,* Thomas.

Lwoff, A. 1950. *Problems of Morphogenesis in Ciliates,* Wiley.

Lwoff, A. and/or S. H. Hutner (Eds). 1951–1964. *Biochemistry and Physiology of Protozoa,* 3 Vols., Academic.

Mackinnon, D. L. and R. S. J. Hawes. 1961. *An Introduction to the Study of Protozoa,* Clarendon.

Sleigh, M. A. 1973. *The Biology of Protozoa,* Elsevier.

Scale, Organization, and Design

2

A. The Size of Ciliates

Ciliates come in a wide assortment of sizes and shapes and even colors, but most are fairly large for single cells. A common ciliate might, for example, be a cylindrical structure with rounded ends, about 100 micrometers long and perhaps 30 micrometers wide. An ordinary prokaryote, such as *E. coli,* might also be cylindrical, with about the same ratio of length to width, but the prokaryote might have a length of only about 1 micrometer. Since the volumes of structures are proportional to the cubes of their linear dimensions, the hundredfold difference in the length of these organisms corresponds to a factor of 10^6 in volume. A paramecium contains enough protoplasm to make approximately 10^6 bacteria.

Ciliates are not large merely in comparison to bacteria. They are large even with reference to the cells of higher organisms. Since an average mammalian cell has a diameter of 5–10 micrometers, our hypothetical ciliate is large enough to contain at least 1000 such cells.

As one might expect from such comparisons, a ciliate is not a "simple" cell with one set of genetic instructions, but rather a "compound" cell, comparable in many respects to a small multicellular animal, such as a rotifer or a nematode. However, it is not divided into cellular compartments. This compoundness without compartmentalization has led to discussions of whether ciliates violate "the cell doctrine" and hence comprise a distinctive kind of "acellular" organism. Such discussions do not touch essential issues, because many organisms at some time depart from the stereotypes of "cellular" organization. Many fungi, for example, become compound by the multiplication of nuclei distributed in a common cytoplasm, even though they are conventionally cellular at other times. Many insects, particularly among the diptera, compound the nuclei in certain tissues by polyploidization or polytenization when this tactic serves a useful physiological or developmental function. The ciliates do provide an opportunity to test how much regional specialization is possible in a large protoplasmic mass that is not subdivided into membrane separated compartments.

B. Nuclear Dimorphism

The ciliate compoundness comes into sharp focus in their unique nuclear organization, which is usually referred to as a state of *nuclear dimorphism*. The manifestations of nuclear dimorphism vary among the ciliates, and these can be arranged in some kind of evolutionary sequence. In the usual situation, however, two kinds of nuclei are present in a common cytoplasm: one or two large nuclei called *macronuclei* and one or a few small nuclei called *micronuclei*. The compoundness of the macronucleus is readily apparent from its DNA content, which varies from perhaps 10 times the amount in the micronucleus in a small ciliate (or even the same amount in the most primitive forms) to 1000 times in a large ciliate.

The two kinds of nuclei have different functions in the cell. The macronucleus controls the continuing metabolic activities of the cell, while the micronucleus serves a reserve function. The evidence for this interpretation comes from various sources. (1) The amount of RNA synthesized per unit of DNA in the macronucleus is very much higher than that synthesized by the micronucleus. At some times in some ciliates no micronuclear RNA synthesis can be detected. (2) Whenever "heterokaryons" have been constructed by placing micronuclei and macronuclei of different genetic constitution in the same cell, gene products of the micronuclear specificity have not been demonstrated. More sensitive tests may eventually provide some evidence for micronuclear function, of course. (3) Some ciliates have no micronuclei, but are nevertheless capable of indefinite multiplication. Such *amicronucleate* cells argue that the functions of the micronuclei are not essential for vegetative processes, but this conclusion cannot be generalized. At least in some ciliates the loss of the micronucleus not only cuts a cell off from a genetic future, but leads to an early somatic death.

Even so the functions of these two kinds of nuclei are very different, and they correspond closely to the functions of two major classes of cells in multicellular forms. The macronuclei are the *somatic* nuclei responsible for vegetative activities of the cell, and the micronuclei are the *germinal* nuclei, which are called upon primarily during sexual episodes, when they produce new somatic nuclei to replace the macronuclei of the previous generation. In agreement with their very different roles, the two kinds of nuclei have different structures, protein compositions, and times of replication. Although the macronuclei are clearly compound, the organization of the chromatin within the nucleus is not fully understood. Indeed, present evidence suggests that different ciliates may have very different macronuclear organization.

C. Cortical Structure

The second distinctive feature of ciliates, and the one responsible for the name of the phylum, is the use of *ciliary units* in the construction of the cellular *cortex*. A ciliary unit is a complex assembly of membranous and fibrillar elements and is capable of considerable variation even within a single species. Characteristically, a ciliary unit contains the elements shown in Figs. 2-1 and 2-2. The *basal body* (or kinetosome) is a cylinder of fibers much like those in the cilium proper, but is missing the central microtubular components, and is located below the surface of the cell. It is continuous with the *ciliary shaft,* which extends beyond the surface of the cell with its 9 + 2 arrangement of microtubules. It is covered by the *ciliary membrane,* which is continuous with the membrane of the cell surface. At regular intervals between ciliary shafts, and in precise geometrical relationships to them, are other

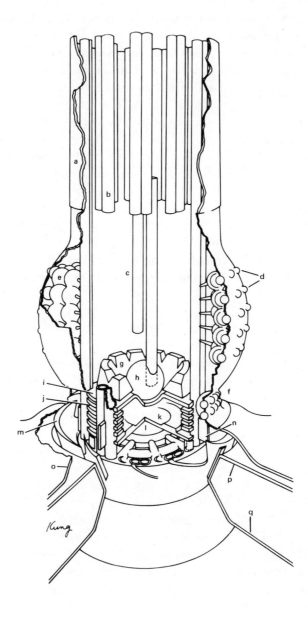

14

membranous and/or fibrous structures, ordinarily located entirely beneath the surface of the cortex, but capable of discharge. These structures in Paramecium are called *trichocysts*. A comparable structure in Tetrahymena is called the *mucocyst* (or mucigenic body). The exact function of these cortical organelles has yet to be discovered; they may play architectural roles, as well as provide protection against certain kinds of hazards. Also characteristically associated with each ciliary unit are one or more fibrous structures that remain permanently beneath the surface. The most general of these is the *kinetodesmal fiber,* which arises near the basal body and projects anteriorly along the right side of the ciliary unit where it eventually joins fibers derived from other basal bodies to compose the *kinetodesma* (pl. kinetodesmata).

Ciliary units in some species have other fibers, originating elsewhere and directed toward certain other structures. In particular, we call attention to the microtubular bands (Fig. 2–2) in Tetrahymena because of their later relevance. A longitudinal set overlies the kinetodesmal fibers; a transverse set arises anterior to the basal body and is directed across the intermeridional space to the cell's left. A third short set of microtubules projects posteriorly and to the right from the posterior right margin of the basal body. The mitochondria may also be considered as

←———————————————————————————

Fig. 2-1. A three-dimensional reconstruction of the proximal region of a Paramecium cilium. (*a*) Ciliary membrane. The outer half of the membrane is lifted in parts to show the necklace and plaque particles. (*b*) Peripheral tubules. Portions of the tubules are deleted to simplify the diagram. (*c*) Central tubules. Note that only one of the tubules enters the axosome. (*d*) Plaque particles. (*e*) Plaque complex. (*f*) Necklace. (*g*) Loosely packed ring of material surrounding the axosome. (*h*) Axosome. (*i*) Curved axosomal plate. (*j*) Rings connecting peripheral tubules. Note that they do not enter the tubules. (*k*) Intermediate plate. (*l*) Terminal plate. (*m*) Transitional fibers (pinwheel structures). (*n*) Projections from the peripheral tubules. (*o*) Plasma membrane. (*p*) Outer alveolar membrane (*q*) Inner alveolar membrane. Reprinted, by permission, from Dute, R. and C. Kung 1978. Ultrastructure of the proximal region of somatic cilia in *Paramecium tetraurelia. J. Cell Biol.,* **78,** 451–464.

Fig. 2-2. A three-dimensional reconstruction of a segment of the cortex of *Tetrahymena pyriformis*. The top of the diagram is directed to the anterior of the cell, and the right side of the diagram (the viewer's right) corresponds to the cell's left, according to the accepted convention. Parts of two ciliary rows are represented with the basal portions of their cilia (C). On the surface midway between the rows are the tips of mucocysts or mucigenic bodies (mb), one of which is shown in a cutaway. Mucocysts are also distributed between the cilia in the ciliary rows. Three prominent bands of microtubules are shown beneath the pellicle. A longitudinal band (lt) extends parallel to the ciliary row on its right. A transverse set (tt) passes from the anterior side of the basal body and across under the longitudinal band before terminating. A short post ciliary set (pt) extends from the longitudinal set to the posterior side of the basal body. At a deeper level than these heavy bands, a pair of basal microtubules (bt) extends along the left margin of each ciliary row. The kinetodesmal fiber (kf) arises at the anterior right margin of the basal body and is directed forward to the right. Mitochondria (M), segments of rough endoplasmic reticulum (er), and ribosomes are also associated with the cortex. Note also the amorphous ectoplasmic material (e). The gaps between the double surface membranes are referred to as alveoli (a). Reprinted, by permission, from Allen, R. D. 1967. Fine structure, reconstruction and possible functions of components of the cortex of *Tetrahymena pyriformis. J. Protozool.*, **14**, 553–565. Fig. 22, p. 564.

structural elements of the cortex insofar as they are imbedded in it in regular patterns.

In the simpler ciliates almost the entire surface of the cell is composed of ciliary units, which may be slightly different in different genera, but which are usually alike within a species. The features we described above show that the ciliary unit is asymmetric, and this asymmetry governs its packing patterns. Each unit has, of course, an inside and an outside. Moreover, the ciliary shaft does not usually emerge directly from the center of the unit, and the kinetodesmal fiber arises from the right side of the unit; these features make it possible to distinguish between the left and right sides of the ciliary unit. The kinetodesmal fiber is directed to the anterior end of the cell and hence indicates an anterior–posterior asymmetry. This anterior-posterior asymmetry is also demonstrated by the place of origin of new basal bodies. New basal bodies usually arise in conjunction with old basal bodies, and particularly in the regions just anterior to them.

Given a building block with a readily identified inside and outside, left and right, and anterior and posterior, one may easily construct a cylinder that approximates the form of a simple ciliate. The packing and growth patterns of the ciliary units give rise to linear arrays of units, called *ciliary rows* (or kineties, or meridians) (Fig. 2-3). The lateral associations between adjacent ciliary rows tend to be more relaxed than those within rows, but they are sufficiently firm to yield a "sheet" of ciliary rows. When the edges of the sheet are brought together, the cylinder is formed, open at each end, but with one end clearly identified as the anterior end by the anterior–posterior asymmetry of the ciliary units.

Even though each ciliary unit has dextrosinistral asymmetry, the closed cylinder does not yet manifest a permanent dorsal-ventral or dextrosinistral bias. This is achieved by the development of a specialized subterminal feeding apparatus, often in association with one specialized ciliary row (the *stomatogenic row* or *kinety number 1*). The feeding apparatus often consists of two special features: one or more *oral membranelles,* composed of

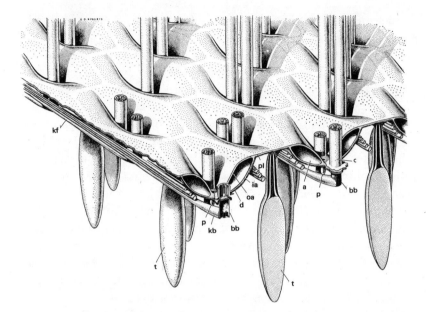

Fig. 2–3. A three-dimensional reconstruction of part of the cortex of *Paramecium aurelia*. The anterior end of the animal is to the reader's upper left, and the ciliary rows are directed diagonally to the reader's left. The kinetodesmal fibers (kf) arise anterior to the basal bodies (bb) and join other fibers running along the (cell's) right margin of the ciliary row. Within the ciliary row the carrot-shaped trichocysts (t) alternate with the cilia (c). Some ciliary units have pairs of cilia. Each cilium has an associated parasomal sac (p). The alveolar space (a) between the ciliary rows is surrounded by an outer alveolar membrane (oa) and an inner alveolar membrane (ia), which is associated with a dense granular layer (d). The microtubular bands are omitted from this diagram. Reprinted, by permission, from Jurand, A. and G. G. Selman, 1969. *The Anatomy of* Paramecium aurelia, Macmillan. Diagram 1, p. 14.

closely packed ciliary units whose ciliary shafts function in coordination to move food materials, and a buccal cavity of specialized membranous and fibrous elements that form *food vacuoles* by a process of *phagocytosis*.

The location of the *oral apparatus* defines the ventral side of the organism, and simultaneously the left, dorsal, and right surfaces of the cylinder. It also serves as the basis for identifying the ciliary rows and the positions of certain other specialized cortical structures. The stomatogenic row is called row 1, and the row to the right of this row is called row 2. By convention, right and left (whether referring to individual ciliary units or to the whole ciliate) are defined from the cell's perspective and not from the perspective of the external observer. One must imagine being inside the ciliate's skin and peering out through the oral apparatus. Row 1 is in the midline, row 2 is slightly to the cell's right, and row number n is just to the left of its midline.

The ciliary rows and the oral apparatus are the major cortical features of a ciliate, but two other structures are almost universal. A *cytoproct,* or cell anus for the discharge of spent food vacuoles, is located in a specific spot on the surface—often near the posterior end of row 1. In addition, most fresh water ciliates possess pulsating water discharge vesicles (*contractile vacuoles*) (Fig. 2–4), which open at specialized places in the cortex called *contractile vacuole pores.* In Paramecium these pores are on the dorsal surface; in Tetrahymena they are on the right side, about one quarter of the distance around the cell.

This simplified introduction to cortical structure will scarcely prepare the beginner to deal with the more complex derivative patterns. Some ciliates, such as the hypotrichs, have no easily identified ciliary rows on their ventral surfaces, but the ciliary units are fused into compound bristles or *cirri* (singular *cirrus*), which serve almost as legs or fins to propel the cells along the surface. In some ciliates, such as Paramecium, the oral apparatus is dislocated from a subterminal position to near the center of the cell. Associated with this change is a complex distortion of ciliary row patterns, so that the primitive anterior–posterior orientation of the rows is difficult to discern. Nevertheless, the basic design elements we discuss above seem to appear with some consistency in the ciliates and can provide a basis for more detailed introduction of particular organisms.

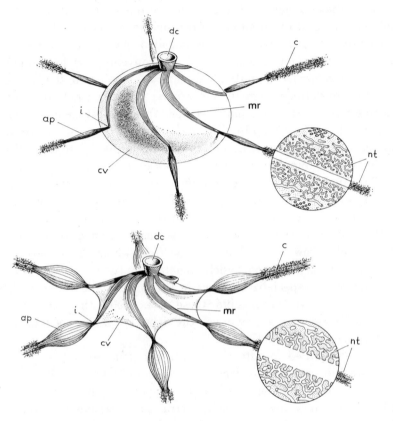

Fig. 2-4. A diagram of the contractile vacuole (cv) apparatus of *Paramecium aurelia* in diastole (top) and systole (bottom). Six nephridial canals (c) are radially disposed around the vacuole. These enter a wider section called the ampullae (ap), connected by a narrow injector canal (i) to the vacuole. Fluids enter the canals through narrow nephridial tubules (nt). Microtubular ribbons (mr) run along the walls of the vacuole to the discharge canal (dc) and contractile vacuole pore. They extend also along the outside of the nephridial canals. Reprinted, by permission, from Jurand, A. and G. G. Selman. 1969. *The Anatomy of* Paramecium aurelia. Macmillan. Diagram 10, p. 43.

D. The Cell Cycle and the Life Cycle

Because of the size and complexity of their cells, the management of the cell cycle poses special problems of integration for the ciliates. Particulate nutrients are directed to the buccal region by ciliary currents. They enter the cell through the formation of food vacuoles in this region. They and all the cytoplasmic contents that are not anchored down are maintained in constant circulation by the patterned flow of the solated cytoplasm called *cyclosis.* Dissolved nutrients can enter the cell in other regions also. Tetrahymena mutants unable to form food vacuoles (Chapter 16) under restrictive conditions can nevertheless grow and divide normally if given an appropriate liquid diet. Phagocytosis is limited to specialized regions, but pinocytosis is a more generalized membrane capability.

The particulate nutrients in the food vacuoles are digested by enzymes secreted into them in an ordered fashion, and patterned changes of pH, associated with the digestive processes, can be followed by feeding the cells indicator dyes. Eventually the unused residues of digestion are egested through the cytoproct.

The nutrients obtained through digestion and pinocytosis provide the raw materials for growth and activity. The proteins and nucleic acids needed to build new cellular fabrics are produced from the raw materials through programmed synthetic patterns. Each phase of the cell cycle is characterized by particular kinds of synthesis and assembly. The DNA for micronuclei and for macronuclei is often synthesized at different times in the schedule. The synthesis, or at least the assembly, of certain cortical components may be restricted to a short interval prior to cell division. Each kind of ciliate has its own particular synthetic program, but in no case do we have an adequate description of the sequence of events, much less a full understanding of the mechanisms. In all cases, however, the program permits the

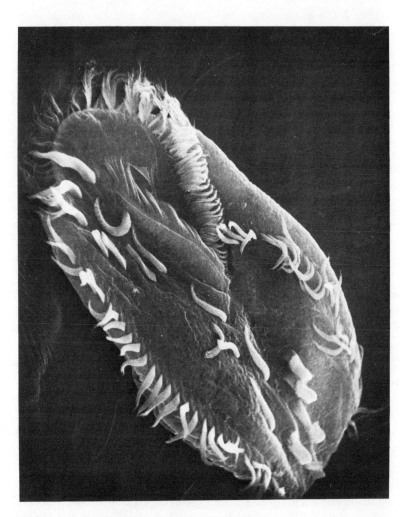

Plate I. Scanning electronmicrograph of *Stylonychia pustulata*. Ventral view of a vegetative cell showing the adoral zone of membranelles, and the ventral and marginal cirri. Magnification: 850×. Courtesy of Gary W. Grimes.

22

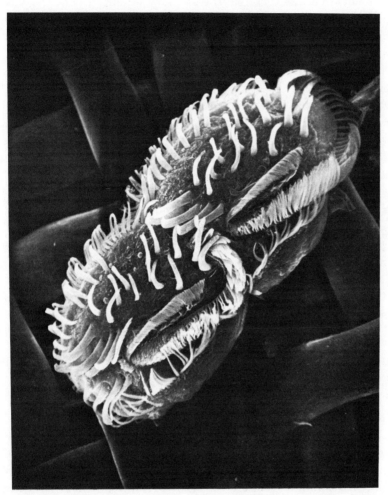

Plate II. Scanning electronmicrograph of *Stylonychia pustulata.* Ventral view of a dividing cell showing the well-developed membranellar and cirral structures on both presumptive daughters. Magnification: 650×. Courtesy of Gary W. Grimes.

stockpiling of components and the insertion of complex organellar elements in a way such that the ordinary running around, feeding, and excreting are minimally disturbed. At the time of cell division all the components essential for two cells have been accumulated, and in a remarkably short time these have been equitably distributed into entirely functional, but approximately half size, organisms.

The necessity for a rapid cytokinesis may provide a partial rationalization of the *amitotic* division of the macronucleus, which leaves each daughter cell with only approximately half the parent's equipment. The significance of this division and its structural implications are considered later (Chapter 8).

This cellular–organismic cycle of synthesis and morphogenesis is repeated many times, but in most organisms it is linked to a program with a longer time span. Ciliates have not only a cell cycle, but also a *clonal life history;* almost imperceptible changes in the synthetic programs lead to distinctive changes in cellular characteristics and eventually (usually) to the death of the cell (Chapter 9) unless some kind of drastic reorganization process intervenes (Chapter 7) to start the program over again.

SUMMARY

The ciliated protozoa represent a unique evolutionary experiment. One of their most distinctive features is their large size, achieved without cellularization, but with the development of a large compound somatic nucleus presiding over a comparably large and complex cytoplasmic organization. By compounding, a ciliate may come to contain a thousand times the protoplasmic mass of one of the simpler eukaryotes.

The compounding of the somatic nucleus is associated with a separation of nuclear function and potentiality. The large somatic nucleus meets most continuing physiological needs, while the almost inert micronucleus bides its time, awaiting another generation. The separation of the germline and the soma in ciliates anticipates the similar strategy developed by multicellular animals.

The large cytoplasmic mass is managed in part by the multiplication of modular structural elements, the ciliary units. The means whereby the complicated architecture is assembled, enlarged, and equitably divided in each cell cycle is a special riddle toward which much study has been directed.

RECOMMENDED READING

Allen, R. D. 1967. Fine structure, reconstruction and possible functions of components of the cortex of *Tetrahymena pyriformis. J. Protozool.,* **14,** 553–565.

Andersen, H. A., L. Rasmussen, and E. Zeuthen. 1975. Cell division and DNA replication in synchronous Tetrahymena cultures. *Curr. Topics Microbiol. Immunol.,* **72,** 1–20.

Ehret, C. F. and E. W. McArdle 1974. The structure of Paramecium as viewed from its constituent levels of organization. In *Paramecium: A Current Survey.* (W. J. van Wagtendonk, Ed.), Elsevier, pp. 263–338.

Hanson, E. D. 1977. The ciliates. *The Origin and Early Evolution of Animals,* Wesleyan University Press, pp. 326–418.

Jurand, A. and G. G. Selman. 1969. *The Anatomy of* Paramecium aurelia, Macmillan.

Orias, E. 1976. Derivation of ciliate architecture from a simple flagellate: An evolutionary model. *Trans. Am. Microsc. Soc.* **95,** 415–429.

The Chosen Few

3

APPROXIMATELY 7200 "NAMED SPECIES" HAVE BEEN assigned to the phylum Ciliata. These have been divided into 3 classes and 23 orders. Because we are interested here in organisms for their utility and promise in experimental biology, we do not need to survey all the diversity available. Indeed, here serious consideration is given to representatives of no more than five genera, and these are unevenly distributed. Both Paramecium and Tetrahymena belong to the order Hymenostomatida (Table 3-1). Both Blepharisma and Stentor are of the order Heterotrichida. Euplotes and most of the other ciliates mentioned in passing belong to the order Hypotrichida. Given below is a brief sketch of some of the important features of the chosen few.

A. Paramecium

Paramecium (Fig. 3-1) is by all odds the most familiar of the ciliates, and it has long been employed in experimental biology. Leeuwenhoek described it with his primitive lenses in the seventeenth century; it was repeatedly observed in the eighteenth century and was a favorite object of study in the Golden Age of Cytology in the late 1800s. Experimental protozoology at the turn of the century provides Jennings' classical description of the "avoiding reaction" in Paramecium (Chapter 15) and the publication of his *Behavior of the Lower Organisms* in 1906. In 1937 T. M. Sonneborn discovered mating types in Paramecium (Chapter 4), the first example in ciliates, and hence was able to bring its sexual processes under control. Paramecium is the most nearly "domesticated" of the ciliates, in the sense that it has been brought under strict laboratory discipline as an instrument for scientific investigation.

The reasons for the popularity of Paramecium are easily understood. Paramecia are large organisms, ranging from 80 to 350 micrometers in length and from 40 to 80 micrometers in width,

29

TABLE 3-1
Classification of Phylum Ciliophora

Phylum CILIOPHORA

Class I. KINETOFRAGMINOPHORA
Subclass (1) Gymnostomata
  Order 1. PRIMOCILIATIDA
      2. KARYORELICTIDA
      3. PROSTOMATIDA
  Suborder (1) Archistomatina
        (2) Prostomatina
        (3) Prorodontina
      4. HAPTORIDA
      5. PLEUROSTOMATIDA
Subclass (2) Vestibulifera
  Order 1. TRICHOSTOMATIDA
  Suborder (1) Trichostomatina
        (2) Blepharocorythina
      2. ENTODINIOMORPHIDA
      3. COLPODIDA
Subclass (3) Hypostomata
  Order 1. SYNHYMENIIDA
      2. NASSULIDA
  Suborder (1) Nassulina
        (2) Microthoracina
      3. CYRTOPHORIDA
  Suborder (1) Chlomydodontina   Chilodonella
        (2) Dysteriina
        (3) Hypocomatina
      4. CHONOTRICHIDA
        (1) Exogemmina
        (2) Cryptogemmina
      5. RHYNCHODIDA
      6. APOSTOMATIDA
        (1) Apostomatina
        (2) Astomatophorina
        (3) Pilisuctorina
Subclass (4) Suctoria
  Order SUCTORIDA
  Suborder (1) Exogenina
        (2) Endogenina
        (3) Evaginogenina

TABLE 3-1 (Continued)

Phylum CILIOPHORA (Continued)	
Class II. OLIGOHYMENOPHORA	
Subclass (1) Hymenostomata	
Order 1. HYMENOSTOMATIDA	
Suborder (1) Tetrahymenina	Tetrahymena
(2) Ophryoglenina	
(3) Peniculina	Paramecium
2. SCUTICOCILIATIDA	
(1) Philasterina	
(2) Pleuronematina	
(3) Thigmotrichina	
3. ASTOMATIDA	
Subclass (2) Peritricha	
Order PERITRICHIDA	
Suborder (1) Sessilina	
(2) Mobilina	
Class III. POLYHYMENOPHORA	
Sublcass Spirotricha	
Order 1. HETEROTRICHIDA	
Suborder (1) Heterotrichina	
(2) Clevelandellina	
(3) Armophorina	
(4) Coliphorina	
(5) Plagiotomina	
(6) Licnophorina	
2. ODONTOSTOMATIDA	
3. OLIGOTRICHIDA	
(1) Oligotrichina	
(2) Tintinnina	
4. HYPOTRICHIDA	
(1) Stichotrichina	Urostyla
(2) Sporadotrichina	Euplotes, Oxytricha, Stylonychia

Source: From Corliss, J. O. 1979. *The Ciliated Protozoa: Characterization, Classification and Guide to the Literature.* 2nd Edition. Pergamon.

31

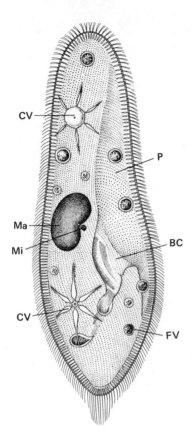

Fig. 3-1. A diagram of *Paramecium caudatum,* indicating the kidney-shaped macronucleus (Ma) and the associated micronucleus (Mi). The two contractile vacuoles (CV) open through pores on the dorsal surface. Cilia in the peristomal groove (P) on the ventral surface direct food particles to the buccal cavity (BC) where food vacuoles (FV) form and circulate around the cell. Reprinted, by permission, from Grell, K. G. 1973. *Protozoology,* Springer-Verlag. Fig. 413, p. 454.

depending on the species. This large size permits microsurgery and electrophysiological measurements on individual cells. The unceasing explorations of paramecia attract the eye and can be followed with low power optics. Their cultivation is simple and inexpensive, and they can be collected in fresh waters the world over.

Paramecium is, of course, a genus, and it contains many species. Even early observers were able to distinguish several major categories of paramecia, particularly the cylindrical or

cigar-shaped *Paramecium caudatum* cluster and the more spatu-
late *Paramecium bursaria* group (Fig. 3–2). Each of these sets,
however, is composed of several morphologically distinguishable
subdivisions. The *P. caudatum* cluster, for example, includes also
Paramecium aurelia and *Paramecium multimicronucleatum*, distinguished on the basis of body size and the form and number of
micronuclei. The *P. bursaria* cluster includes not only the green *P.
bursarias* (carrying photosynthetic chlorellae), but transparent
forms such as *Paramecium trichium* and *Paramecium calkinsi.* As
we see later, many, and perhaps all, of these morphotypic
"species" are in fact complexes of noninterbreeding "genetic
species." Sonneborn has identified and named some 14 species in
the *P. aurelia* complex. Studies have been carried out on many of
these species, but two species have been particularly important:
Paramecium primaurelia and *Paramecium tetraurelia.*

The "domestication" of an organism for investigative biology
requires particularly that its growth and its sex life be brought
under control, for certain research strategies require a combined
genetic and biochemical technology. Its ease of cultivation has
made Paramecium a popular subject of investigation, but its
nutritional requirements have presented problems. In nature
paramecia feed on bacteria, and they are ordinarily cultivated in
the laboratory on bacterized media, usually with a controlled
pedigreed strain of *Klebsiella pneumoniae.* These "monoxenic"
cultures permit good and reproducible growth, but the medium is
not "defined." For many purposes the presence of bacteria (and
plant extracts) is unimportant, but biochemical studies are sometimes compromised. Axenic and defined media are now available,
however, and these facilitate coordinated biochemical and genetic
investigations.

The genetic technologies with the *P. aurelia* species are aided
greatly by the regular occurrence of autogamy. Autogamy is a
process whereby the micronucleus (1) becomes homozygous and
(2) generates a homozygous macronucleus (Chapter 7). A mutagenized culture may be induced en masse to undergo autogamy,

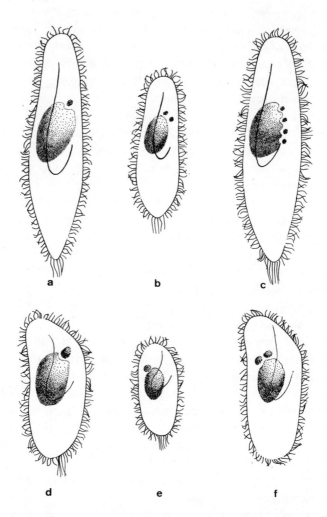

Fig. 3-2. Outlines of some common named species of Paramecium sketched to the same scale. (a) *P. caudatum,* (b) *P. aurelia,* (c) *P. multimicronucleatum,* (d) *P. bursaria,* (e) *P. trichium,* (f) *P. calkinsi.* Some protozoologists believe the last three "species" are sufficiently distinct to be placed in a different genus. Reprinted, by permission, from Vivier, E. 1974. Morphology, taxonomy and general biology of the genus Paramecium. In *Paramecium, A Current Survey* (W. J. van Wagtendonk, Ed.), Elsevier. Fig. 23, p. 34.

34

and the exautogamous population may be screened directly for mutants. The ease of inducing and selecting mutations in an autogamous species is comparable to that in a haploid organism and is much greater than in most diploid organisms, particularly crossbreeding diploid organisms.

One somewhat disadvantageous feature of the *P. aurelia* complex is the number of chromosomes. Certain kinds of genetic studies are facilitated by well-defined linkage maps, and the ease of developing linkage groups is related to the number of chromosome pairs. The *P. aurelias* are somewhat diversified karyotypically, even within a species, and the chromosomes are small and numerous, ranging from about 40 to 50 pairs in the strains studied.

B. Tetrahymena

Although Paramecium is the most familiar of the experimental ciliates, Tetrahymena (Fig. 3-3) is the most important recent addition to the category. It was certainly seen in the nineteenth century, but its small size (40–60 micrometers) and generalized form prevented its ready discrimination. Its first claim to attention arose from André Lwoff's establishment of an axenic culture in 1923. The strain he isolated is still maintained in many laboratories. Although, like Paramecium, Tetrahymena in nature is generally a bacterial feeder, Tetrahymena is much easier to cultivate on artificial media; it grows well, for example, on 1% proteose peptone. It can be enriched easily from natural waters by the use of proteose peptone and antibiotics (to eliminate bacteria and obligate bacterial feeders). It was the first ciliate to be grown on defined media, through the intensive studies of Dewey and Kidder beginning in the 1940s, and has been the subject of numerous nutritional and pharmacological studies since that time. (A characteristic growth medium is shown in Table 3-2.) The similarity between ciliate nutritional requirements and those of

TABLE 3-2

**Medium for growing *Tetrahymena pyriformis*;
weight of components for 100 ml of medium**

Medium	Wt. for 100 ml of final medium (mg)
Amino Acids	
DL-Alanine	30
L-Arginine · HCl	30
L-Asparagine · HCl	20
L-Glutamic acid	40
L-Glutamine	10
Glycine	40
L-Histidine · HCl · H$_2$O	20
L-Isoleucine	20
L-Leucine	20
L-Lysine · HCl	20
DL-Methionine	30
DL-Phenylalanine	30
L-Proline	20
DL-Serine	30
DL-Threonine	40
L-Tryptophan	15
DL-Valine	20
Nucleosides	
Adenosine	2
Cytidine	2
Guanosine	2
Uridine	2
Carbohydrates	
Glucose	1000
Salts	mg
K$_2$HPO$_4$	25
KH$_2$PO$_4$	25
MgSO$_4$ · 7H$_2$O	50
CaCl$_2$	1
Citric acid	60
Vitamins	
Nicotinic acid	90
d-Pantothenate · Ca	75
Thiamine · HCl	50

TABLE 3-2 (Continued)

Medium	Wt. for 100 ml of final medium (mg)
Riboflavin-5′-phosphate · Na	45
Pyridoxamine · 2HCl	5
Pyridoxal · HCl	5
Biotin	0.1
DL-6-Thioctic acid	1
Folinic acid, calcium salt	1
Trace metals	
$Fe(NH_4)_2(SO_4)_2 \cdot 6H_2O$	1400
$ZnSO_4 \cdot 7H_2O$	450
$MnSO_4 \cdot 4H_2O$	160
$CuSO_4 \cdot 5H_2O$	30
$Co(NO_3)_2 \cdot 6H_2O$	50
$(NH_4)_6Mo_7O_{24} \; 4H_2O$	10

Source: From Rasmussen L. and L. Modeweg-Hansen 1973. Cell multiplication in Tetrahymena cultures after addition of particulate material. *J. Cell Sci.,* **12,** 275–286.

mammals provided a (probably mistaken) suggestion of close affinity (Chapter 5).

The characterization of Tetrahymena, unlike that of Paramecium, came quite late, after improved silver impregnation techniques permitted discrimination among the many small ciliates. The name Tetrahymena, referring to the four membranelles of the oral apparatus, was first applied in 1940 by Furgason, who provided the means of distinguishing it from such forms as Glaucoma and Colpidium. The earlier literature on these small ciliates is almost hopelessly confused because identification was so difficult. Corliss's studies from the 1950s provide for a reasonably satisfactory assortment of these smaller organisms.

Morphological studies, however, do not reveal the genetic and evolutionary relationships of the Tetrahymenas. As in the case of Paramecium, the morphotypic species of Tetrahymena consist of

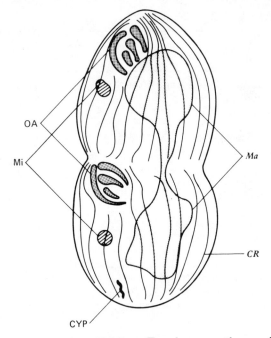

Fig. 3-3. Diagram of a dividing *Tetrahymena thermophila* at the "peanut stage." The longitudinal lines represent the ciliary rows (CR). The micronuclei (Mi) have already divided. The macronucleus (Ma) is constricted prior to its amitotic division. A new oral apparatus (OA) has been developed for the posterior daughter; the anterior daughter retains the old oral apparatus. Note the four oral membranelles from which Tetrahymena received its name. The cytoproct (CYP) or cell anus remains with the posterior daughter and a new one is developed later for the anterior daughter.

multiple genetic species. Again as in the case of *Paramecium aurelia,* when identification of species without the use of living reference strains became possible, Latin binomials were applied to the genetic species. Species discrimination within the species complex of *Tetrahymena pyriformis* is easily achieved so long as molecular characteristics are examined, but the organisms are remarkably alike in size, architecture, nutritional requirements, and karyotypes. The discrepancy between the phenetic and genetic dispersion of these species, so much more striking than in the *P. aurelia* complex, is a special challenge to the evolutionist (Chapter 5).

Although the culture of Tetrahymena and nutritional studies ran ahead of those in Paramecium, genetic analysis lagged considerably behind. Mating types in Tetrahymena were discovered by Elliott and Gruchy only in 1952, and genetic studies were slow gathering momentum. *Tetrahymena thermophila,* on which genetic studies were concentrated, is an outbreeding species—like corn and mice and men—and showed considerable inbreeding depression when attempts were made to develop homozygous strains. Moreover, clones maintained in laboratory culture underwent a kind of senescence and gradually lost the ability to produce viable progeny at conjugation, so that frequent crosses were necessary to maintain stocks (Chapter 9). Only in the 1960s was the problem of preservation of germplasm solved through the use of liquid nitrogen storage.

Even so, the perfecting of this organism for genetic analysis required an economical means of inducing and selecting mutants. The absence of autogamy was sorely felt. Gradually techniques for inducing homozygosity in mutagenized stocks were developed— first genomic exclusion by Allen, then short circuit genomic exclusion by Bruns, and finally cytogamy induced by hyperosmotic shock by Orias and Hamilton. The details of these processes and their employment in investigations are treated later (Chapter 7). For present purposes it is necessary only to state that the genetic capabilities of Tetrahymena at last approach those of Paramecium. One genetic advantage of Tetrahymena appears to be the low number of chromosomes—five pairs in all the strains thus far examined.

Note should be taken of the fact that not all Tetrahymena species are sexual, and ordinary genetic analysis is limited to those with micronuclei that can be induced to conjugate. Strain GL, the primordial Tetrahymena isolated by Lwoff, is an amicronucleate strain assigned to the species *T. pyriformis (sensu stricto),* whose genetics is limited to the analysis of macronuclear mutations. Unfortunately, much of our earlier knowledge concerning tetrahymenas comes from these asexual strains, and that knowledge may need to be confirmed for the breeding species before certain coordinated genetic–biochemical studies can be undertaken with confidence.

C. Stentor

Stentors (Fig. 3–4) are even larger than paramecia and are commonly observed in hay infusions as active and colorful animals. When extended some Stentors reach a millimeter in length. They early attracted the attention of microsurgeons and have proved to be remarkably tolerant to vivisection. They have been exploited in developmental studies by several gifted students, notably Weisz, Tartar, Uhlig, and de Terra.

Unfortunately, in contrast to Paramecium and Tetrahymena, Stentor is scarcely domesticated at all. Stentors are still grown in mixed cultures of microorganisms and are periodically re-collected from natural sources. Standard lineages are not maintained and identification is made only to one of the five morphotypic species. In view of the underclassification demonstrated with other ciliates, even these designations must be suspect.

The laboratory regulation of Stentor's sex life is also far behind that in other ciliates. Conjugation is occasionally observed in cultures, but reliable means of inducing mating have not been reported. Consequently, essentially nothing is known of the genetics of these organisms, and the prospects of combined genetic and biochemical analyses are remote.

Nevertheless, these organisms have contributed substantially to our understanding of the means whereby protoplasm organizes its components, and they provide answers to some questions not easily posed to their smaller, if more sophisticated, cousins (Chapter 10).

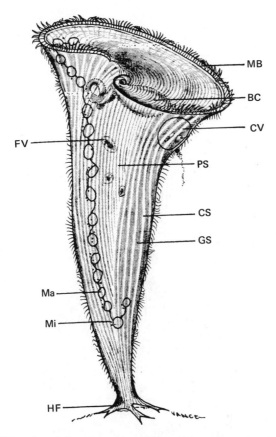

Fig. 3–4. Diagram of the structure of *Stentor coeruleus.* A membranellar band (MB) encircles the anterior end and brings food particles to the buccal cavity (BC), where food vacuoles (FV) are formed. The ciliary rows extend from the holdfast structure (HF) to the membranellar band, and this longitudinal organization is made visible by the presence of pigmented granules in the cortex arranged to make contrasting granular stripes (GS) and clear stripes (CS). Another cortical contrast is provided by the differences in the widths of the stripes. The ventral narrow stripes become progressively wider to the cell's right, and a region of discontinuity appears where the narrow and broad stripes abut. This region of stripe contrast is the primordium site (PS) in which the new oral primordium arises at cell division. The contractile vacuole (CV) has a characteristic location with reference to the cortex. The vegetative macronucleus (Ma) takes the form of a chain of macronuclear nodes. The micronuclei (Mi) appear as numerous small black spots along the macronuclear chain. Reprinted, by permission, from Tartar, V. 1961. *The Biology of Stentor,* Pergamon. Fig. 1, p. 8.

41

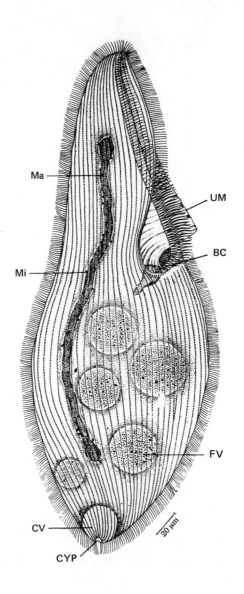

42

D. Blepharisma

Blepharisma (Fig. 3–5) is another large, familiar, common ciliate that has not been significantly domesticated, but its laboratory manipulation is somewhat advanced over that of Stentor. Several different species have been described. All are elongated flexible organisms; some have a pink pigment that is responsible for their "photodynamic" properties and that accounts for much of the physiological work that has been done on them.

The culture methods for Blepharisma are still primitive; the organisms are characteristically grown on bacterized lettuce infusion, like Paramecium. The most intensively studied forms are of the species *Blepharisma japonicum.* In addition to the usual pink form, a so-called albino variant has been maintained for many years, and this strain has been useful in the analysis of cellular interactions.

Very little genetic work has been done on Blepharisma, and only recently has mating been brought under control. Pairs form in unmixed cultures, and for many years mating types were thought not to exist. Recently, however, Miyake and his collaborators have shown the existence of two mating types and have explored carefully the circumstances of their interaction. Ironically, Blepharisma is the only ciliate for which a significant molecular description of mating has been begun. The characteristic soluble pheromones of the mating types have been isolated, characterized, and, in one case, synthesized (Chapter 4).

←————————————————————————————

Fig. 3–5. Diagram of *Blepharisma japonicum.* An undulating membrane (UM) directs food particles to the buccal cavity (BC) on the ventral surface, where food vacuoles (FV) develop. Wastes are egested at the cytoproct (CYP) at the posterior end; the contractile vacuole (CV) is located in the same region. The vegetative macronucleus (Ma) is a flexible elongated rod running most of the length of the cell and associated with numerous small micronuclei (*Mi*). Reprinted from *Blepharisma: The Biology of a Light-Sensitive Protozoan,* by Giese, Arthur C. with the permission of the publishers, Stanford University Press. © 1973 by the Board of Trustees of the Leland Stanford Junior University. Fig. 1.2, p. 7.

E. Euplotes

Choosing a fifth ciliate genus for special consideration is difficult, and somewhat arbitrary. We are trying to select candidates for their present utility rather than for their potentiality or as representatives. Paramecium and Tetrahymena are both holotrichs. Stentor and Blepharisma are both members of the spirotrich order. Although 23 orders of ciliates are presently identified (Table 3-1), we obviously cannot include more than three orders among our chosen five genera. The arbitrariness in the choice of the fifth team member refers more to the particular genus chosen than to the order. The large, highly specialized much studied order of hypotrichs must be the choice. The determination of the most appropriate genus is problematical because the experimental work has been spread fairly evenly among several similar genera: Euplotes, Stylonychia, Oxytricha, Urostyla, and so on. Euplotes is chosen primarily because earlier and more extensive and systematic genetic work has been carried out on it than on the others.

Euplotes (Fig. 3-6), like the other members of the order, is flattened dorsoventrally and has very different ciliature on the two surfaces. Particularly, as is mentioned above, the ventral surface bears fused ciliary organelles called *cirri,* which are used almost as legs for walking on the substratum. The dorsal surface bears rows of short and separated cilia. Much study has been directed to the morphogenesis of these structures.

The culture of the hypotrichs is still a major problem. They are ordinarily grown on bacterized media; sometimes tetrahymenas are supplied as food organisms.

Breeding studies have been extensive among the hypotrichs, but they have been scattered over several species. The morphotypic species are often composed of several cryptic "genetic species." Each species commonly has several mating types, sometimes many. The nuclear behavior has been studied and genetic markers have been followed. No substantial body of genetic information

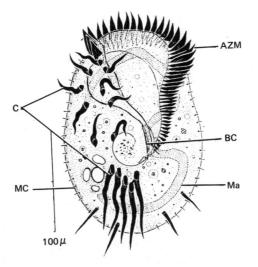

Fig. 3-6. Diagram of the structure of Euplotes. The large adoral zone of membranelles (AZM) again directs particulate food to the buccal cavity (BC). The vegetative macronucleus (Ma) is a long contorted rod; the micronuclei are inconspicuous. The cilia of the ventral surface are compounded into distinctive cirri (C) whose numbers and positions are precisely prescribed. Rows of compound cilia are also distributed along the edges of the flattened cell as marginal cirri (MC). Reprinted, by permission, from Dragesco, J. 1970. Ciliés *libres du Cameroun. Ann. Fac. Sci. Yaoundé, Univ. Fed. Cameroun.,* 1–141, Fig. A, p. 130.

has been assembled for any one species, however, and few well-marked breeding strains are maintained. Mutational dissection of cellular characteristics has not been attempted, though adventitious mutants or strain variants have been employed effectively.

Again, somewhat ironically, these organisms so poorly domesticated have nevertheless been the source of important information in comparative molecular studies, particularly with respect to macronuclear organization (Chapter 8).

SUMMARY

Five groups of related species provide the basis for most modern experimental work on the ciliates. The genus Paramecium is the most familiar group and the earliest studied. It is the most nearly "domesticated" as a scientific research tool, unless the genus Tetrahymena has overtaken it. The tetrahymenas have a disadvantage in being smaller, but are more easily cultured in defined media.

None of the other ciliates is yet approaching domestication, but several are useful in comparative studies. The genus Stentor has been exploited for its large size and ability to tolerate surgical reconstruction; it has provided much of our understanding of ciliate architectural dynamics. Blepharisma, in contrast, is providing special insight into the kinds of cellular signals that coordinate behavior within and between individuals of a species. Finally, the hypotrichs provide several large, complex, and attractive organisms whose genetic and molecular organization is being rapidly unraveled.

GENERAL WORKS CONCERNED WITH THE CHOSEN FEW

Beale, G. H. 1954. *Genetics of* Paramecium aurelia, Cambridge University Press.

Elliott, A. M. (Ed.) 1973. *Biology of Tetrahymena,* Dowden, Hutchinson, and Ross.

Giese, A. C. 1973. *Blepharisma: The Biology of a Light-sensitive Protozoan,* Stanford.

Hill, D. L. 1972. *The Biochemistry and Physiology of Tetrahymena,* Academic.

King, R. C. (Ed.) 1975. *Handbook of Genetics,* Vol. 2, Plenum.

Tartar, V. 1961. *The Biology of Stentor,* Pergamon.

van Wagtendonk, W. J. (Ed.) 1974. *Paramecium: A Current Survey,* Elsevier.

Wichterman, R. 1953. *The Biology of Paramecium,* Blakiston.

Mating, Mating Types, and Species Complexes

4

A. Mating Systems

Conjugation was observed in ciliates very early in field collections and laboratory cultures and was eventually interpreted as a sexual event, but it was not brought under control immediately and hence was not made into a useful tool. The discovery of *mating types* in Paramecium changed this situation. Sonneborn showed that deliberate mixtures of clones, even clones derived (by autogamy) from the same ancestral cell, would sometimes yield massive agglutination, followed by cell-to-cell pairing. Since intraclonal mating does not usually occur, but interclonal mating does, the cells of different origin obviously perceive a difference that is not apparent to the human observer. These differences are referred to as *mating type* differences and are ordinarily defined by the use of live *reference strains.* The term mating type is preferable to *sexes,* because sexes suggests a differentiation of gametic function that is inapplicable in conjugation. In most ciliates the cells mating are not distinguishable in form or function; conjugation is an act of double cross-fertilization so that cells of any mating type are in a sense hermaphroditic. Another reason for using the term mating types instead of sexes is that in some species the number of complementary mating types is very large, and the bipolar vocabulary of "sexuality" is inadequate to cope with it.

The discovery of mating types provided a powerful tool for investigating the relationships among organisms of similar appearance. Sonneborn found that strains of paramecia conforming to the general description of *P. aurelia,* consist, in fact, of many different sets of mating types. Mating types I and II mate with each other and mating types V and VI mate together (Table 4-1). But no mating occurs when the other combinations of these types are made. Strains of types I and II are genetically isolated from strains V and VI. Therefore strains of types I and II constitute a *genetic species* distinct from that made up of strains of types V and VI.

49

TABLE 4-1

Mating reactions among 8 of the 14 species of the *P. aurelia* complex

The species here are designated by numbers, referring back to the time they were called "syngens." Species 1 is now called *P. primaurelia*; species 4 is *P. tetraurelia*. The species are separated into two "groups," depending on their mode of mating type determination. Group A species have random karyonidal determination (Chapter 13); group B species have coordinated karyonidal determination (Chapter 14). Specific mating types are indicated by roman numerals (I and II, VII and VIII), but homologies among them allow assignment to Even and Odd general categories. Interspecific conjugation occurs in some combinations, but leads invariably to death in F_1 or F_2. In some combinations firm pairs do not form, but tentative (t) interactions occur in appropriate combinations. The numbers indicate the intensity of mating. Intraspecific conjugation commonly yields 95% or more of the cells in pairs.

Group		Species	Mating Type	I	II	V	VI	IX	X	VII	VIII	XIII	XIV	XV	XVI	XIX	XX	XXIII	XXIV	Type
A		1	I	0	95	0	40	0	0	0	0	0	0	0	†	0	0	0	0	Odd
			II		0	1	0	40	0	0	0	10	0	†	0	0	0	0	0	Even
		3	V			0	95	0	0	0	0	0	0	0	40	0	0	0	0	Odd
			VI				0	0	0	0	0	†	0	0	0	0	0	0	0	Even
		5	IX					0	95	0	0	0	0	0	0	0	0	0	0	Odd
			X						0	0	0	†	0	0	0	0	0	0	0	Even
B		4	VII							0	95	0	0	0	95	0	0	0	0	Odd
			VIII								0	0	0	60	0	†	0	†	0	Even
		7	XIII									0	95	0	†	0	0	0	0	Odd
			XIV										0	0	0	0	0	0	0	Even
		8	XV											0	95	0	0	0	0	Odd
			XVI												0	†	0	†	0	Even
		10	XIX													0	95	0	0	Odd
			XX														0	†	0	Even
		12	XXIII															0	95	Odd
			XXIV																0	Even

Source: From Butzel, H. M. 1974. Mating type determination and development in *Paramecium aurelia.* In *Paramecium, A Current Survey* (W. J. van Wagtendonk, Ed.), Elsevier, Fig. 1, p. 92.

A total of 14 such genetic species have been described in the *P. aurelia* complex and each of these species has two mating types. The species were originally designated as *varieties* and later still were called *syngens*. Giving the species Latin binomials was resisted, mainly on pragmatic grounds. The species of the complex were so similar that reliable discrimination required living reference strains. A diagnostic system requiring such strains was considered insecure, for if the reference strains were lost, the identity of new strains would be difficult or impossible to determine. Recently, however, the species have been found to be identifiable by isozyme techniques that do not require reference strains. Therefore Latin binomials have been assigned to the species of the *P. aurelia* complex. To retain some continuity of designation, the new names echo the earlier varietal or syngenic names. Syngen 1 has become *P. primaurelia*, syngen 2 is now *P. biaurelia*, and so on.

Although each species of the *P. aurelia* complex is completely isolated from the gene pools of the other species, some signs of interaction between species are seen (Table 4-1). In some cases weak agglutination is found with interspecific mixtures; in some cases pairs are formed, but the F_1 is inviable. In one instance pairing is nearly normal and the immediate offspring are vigorous, but all the F_2 and backcross progeny die.

Although gene pool isolation is complete, the interspecific mating reactions establish the homologies of the mating types in the several species. In no instance does one mating type of one species conjugate with *both* mating types of another species. If we arbitrarily call mating type I of *P. primaurelia* the odd mating type, and type II the Even mating type, we may assign Even and Odd designations to the mating types of most of the species on the basis of the interspecific interactions. It is never necessary to postulate a mating between two Odd cultures or two Even cultures. The homologies suggested by such analyses are supported by other evidence as we see later (Chapter 13).

Many, perhaps most, of the originally described ciliate species are actually species complexes—though relatively few have been fully analyzed. Those that have been examined, however, show that the features described for the *P. aurelia* complex are not universal. The *P. caudatum* and *P. multimicronucleatum* complexes are similar to that of *P. aurelia,* in that each of the species (not yet given Latin binomials) has two mating types. Such systems may be designated as *binary systems.* In contrast, the *P. bursaria* complex consists of several species—each of which contains not two distinctive mating types, but either four or eight (Table 4-2). The mating game within these species permits conjugation between cells of any two different types.

TABLE 4-2

The system of mating reactions within two of the genetic species (syngens) of the *P. bursaria* complex

Unlike the species of the *P. aurelia* complex, each species has either four or eight mating types. Conjugation within a species occurs between any clones of unlike type, but conjugation does not occur between species.

Species		Mating Type													
		A	B	C	D	E	F	G	H	I	J	K	L	. . .
1	A	−	+	+	+	−	−	−	−	−	−	−	−	
	B	+	−	+	+	−	−	−	−	−	−	−	−	
	C	+	+	−	+	−	−	−	−	−	−	−	−	
	D	+	+	+	−	−	−	−	−	−	−	−	−	
2	E	−	−	−	−	−	+	+	+	+	+	+	+	
	F	−	−	−	−	+	−	+	+	+	+	+	+	
	G	−	−	−	−	+	+	−	+	+	+	+	+	
	H	−	−	−	−	+	+	+	−	+	+	+	+	
	I	−	−	−	−	+	+	+	+	−	+	+	+	
	J	−	−	−	−	+	+	+	+	+	−	+	+	
	K	−	−	−	−	+	+	+	+	+	+	−	+	
	L	−	−	−	−	+	+	+	+	+	+	+	−	
.														
.														
.														

The simplest interpretation of *multiple systems* is that they have arisen by the duplication of a primitive binary system. If a binary system is described as being composed of cells with one of two complementary surface properties A or α, then a duplicated system might involve A or α, *and* B or β; a clone could be described as AB, Aβ, αB, or $\alpha\beta$. Clones differing in *either* or *both* complementary systems would be able to mate. The fact that all the mating systems studied in the genus Paramecium have numbers of mating types conforming to a 2^n series, strongly supports the idea that a basic binary complementation mechanism has been compounded to yield the various multiple systems in the different species.

Other ciliate "taxonomic species" outside the genus Paramecium also consist of sibling species complexes. Some of these species have binary mating systems and others have multiple systems, but the relationship between the mating mechanisms in these organisms and that in paramecia is uncertain. For one thing, the numbers of mating types in some of these other species may be very large; 48 have been reported in one species of Stylonychia, and the numbers do not obviously follow the 2^n series found in Paramecium. Some of the mating systems involve soluble pheromones, or *"gamones,"* unlike the Paramecium mating substances that appear to be membrane bound at all times. The genetic controls of the mating types may also be very different in the various species. For such reasons, a satisfying synthetic view of ciliate systems must be deferred to a later date.

B. Social Interactions

Mating is the chief, perhaps the sole, social event in the life history of most ciliates. We have little evidence for cooperative interactions, as in feeding or resisting predators, or for territorial antagonisms, though spacing mechanisms may exist in sessile forms. Generally, ciliates seem to go their own individualistic ways and come together only for the object of sexual recombination. These sexual episodes do not occur at random, however, but are rigorously regulated as to the intervals between matings and the participants in the interactions (Chapter 6). The mechanisms whereby ciliates recognize appropriate mating partners and arrange genetic junctions have been explored in several species, but they are still incompletely understood.

The cells mating must ordinarily be of the same "genetic species" and of different mating types. But not all conspecific mixtures of different mating types yield conjugants. Several other conditions must be met. In many genera the cells that unite in conjugation must be moderately starved and in the same phase of the cell cycle. The nutritional requirement is easily rationalized. A severely starved cell lacks the nutritional reserves to support the metabolic demands of sexual reorganization. On the other hand, a ciliate that mates when food is available is likely to lose in competition with conspecific cousins that mate only when food supply is exhausted. Conjugation is a complex reorganization process that requires many hours to complete; in some species several days are required. Mating is not a multiplicative process, but usually begins and ends with two cells. A mating cell of *P. tetraurelia* requires about 24 hours to conjugate, reorganize, and undergo its first two postzygotic divisions. In the same interval a nonmating cell usually undergoes 5 cell divisions and is represented by 32 cells instead of 4. In *T. thermophila*, a vegetative cell can undergo eight divisions and produce 256 cells in the time required for a complete conjugation sequence. The selective disadvantage of mating when food is at hand is an intolerable burden.

The prescription that mating should occur only when the cells are moderately starved is also rationalized on another basis. Cells in fast exponential growth are unsynchronized with respect to the cell cycle, and a chance association of two cells would likely pair cells in different physiological states. Conjugation is a complex sequence of events requiring careful coordination of events in the mating partners. Both pair members must undergo two meiotic divisions, and one or more additional prezygotic nuclear divisions, and prepare to exchange a pronucleus with the mate at some precise time several hours after the beginning of the process. The temporal programs in the mating cells must proceed in parallel. Although some evidence is available for cross monitoring, having the cells in equivalent physiological and cell cycle states at the beginning is important in assuring temporal compatibility. Indeed, in any given species the cells that start mating are usually at the same point in the cell cycle, though the "at-rest" state for different species may be different.

Having cells in the appropriate nutritional condition again is not sufficient to assure mating. Conjugation is a highly specialized maneuver, requiring processes and products unique to this point in the life cycle. The specialized syntheses needed for conjugation would be wasted in organisms reaching the limits of their food supply in the absence of suitable mates. Cells do not fully prepare to mate unless they are assured of the availability of mates in the vicinity. This assurance can come either through direct contact with potential partners or through molecular signals transmitted from a distance (*gamones*). In either case the cell must produce signals and must have sensors for the signals from potential mates.

The molecules signaling mating availability in Paramecium and Tetrahymena are firmly bound to the surface membranes, but differ in their mode of action. In Paramecium the first contact between suitable mates results in ciliary agglutination, which serves to retain the cells in the same vicinity while the special substances required for mating are being synthesized. Tetrahymena, on the other hand, show fewer overt signs of the information exchange that occurs during a collision of potential mates; but the change of synthetic program is begun, and it is perhaps sustained by repeated collisions through an interval of "co-stimulation" (Fig. 4-1). At the end of an hour or so the potential mates are prepared to enter into a more intimate relationship.

The mating act in ciliates requires more than a random fusion

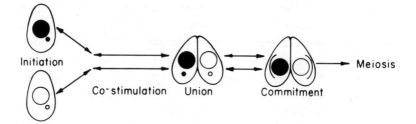

Fig. 4-1. Summary of early events in Tetrahymena conjugation. The process of *initiation* requires the starvation of mature cells, but it may occur in unmixed cultures. Initiated cells become capable of cell union only through the process of *co-stimulation,* which requires the physical presence of cells of complementary type. Cell union may be reversed in early stages by the addition of nutrient, but only until the time the cells become *committed* to meiosis. Reprinted, by permission, from Nanney, D. L. 1977. Cell-cell interactions in ciliates: Evolutionary and genetic constraints. In *Microbial Interactions* Series B, Vol. 3, (J. Reissig, Ed.), Chapman and Hall. Fig. 9.3, p. 376.

of cell surfaces, for the cell body is large and the small nuclei have to be precisely positioned. In both Paramecium and Tetrahymena (and indeed in all ciliates probably) a proper "docking" of the mates is essential. Whenever an inappropriate orientation occurs, the subsequent behavior of the nuclei is disturbed. In the *Paramecium aurelia* group, during the agglutinative co-stimulating stage the cells develop specialized adhesive surfaces on the ventral side and lose the surface organelles at those sites. One surface at the anterior end is called the holdfast region; one in the midregion, through which the pronuclei are eventually transferred, is called the paroral cone region. A third point of attachment further posterior is also sometimes observed. With the acquisition of a hold-fast attachment, the mating cells lose their general ciliary agglutinability and are committed to conjugation.

In Tetrahymena, perhaps because of its smaller size, only one mating surface is present, anterior to the oral apparatus. The haploid nucleus destined to produce the pronuclei moves to a special point in the attachment area. To avoid a nuclear collision during

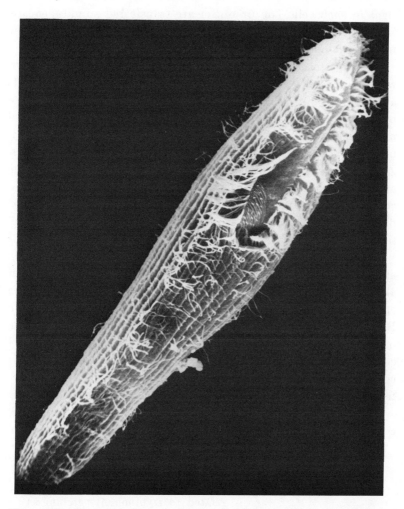

Plate III. Scanning electronmicrograph of *Blepharisma sp.* Published
with permission from Small, E. B., D. S. Marszalek, and G. A. Antipa
1971. A survey of ciliate surface patterns and organelles as revealed with
scanning electron microscopy. Magnification: 740✕. *Trans. Am. Microsc.
Soc.*, **90,** 283–294.

transfer, the nuclear attachment point is displaced slightly to the right in each cell. The pronuclei are transferred simultaneously. Curiously enough, as Maupas long ago observed, all ciliates obey the right-hand rule of nuclear passage.

Blepharisma manages these matters in a somewhat different way. It does not rely on random cellular contacts to signal its availability, but secretes a soluble product that serves as a messenger. Each of the two mating types produces such a chemical messenger, or gamone, but the interactions are not entirely symmetrical. The gamone of mating type II (blepharismone, or gamone 2) (Fig. 4-2) is a small molecule that is freely diffusible and that serves as a chemoattractant of type I cells. Gamone 1, in contrast, is a large molecule, a glycoprotein of about 20,000 daltons (excluding tryptophan); it is not apparently chemoattractive.

In other respects, the effects of the two gamones are much alike. Each triggers in the target cells a readiness to mate, brought about by the induction of specialized protein synthesis and a change in the surface membranes (Fig. 4-3). The agglutination in Blepharisma differs from that in Paramecium and Tetrahymena, in that activated cells stick together more or less at random; type I cells may form pairs with type I cells as readily as with type II cells. In Paramecium and Tetrahymena mating cells are of different mating types except under unusual circumstances.

However, homotypic pairs in Blepharisma do not complete conjugation. They may remain together for very long periods of

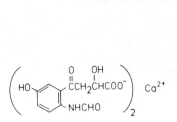

Fig. 4-2. The chemical structure of blepharismone, gamone II, a small molecule related to tryptophan: calcium-3-(2´-formylamino-5´ hydroxybenzol) lactate. Reprinted with permission from Miyake, A. 1978. Cell communication, cell union, and initiation of meiosis in ciliate conjugation. *Curr. Topics Dev. Biol.*, **12,** 37–82. Fig. 4, p. 41. Copyright by Academic Press Inc. (London) Ltd.

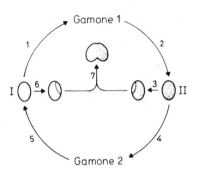

Fig. 4–3. Cell interactions in early conjugation in *Blepharisma japonicum.* Gamone 1, produced by type I cells, activates cells of type II and stimulates them to increase the production of gamone 2 which has a symmetrical effect on cells of type I. Gamone 2 is also a chemoattractant for cells of type I. Reprinted, by permission, from Miyake, A. 1978. *Curr. Topics Dev. Biol.,* **12,** 37–82. Fig. 3, p. 41. Copyright by Academic Press Inc. (London) Ltd.

time, so long as their reactivity is sustained by the presence of the gamones, but they do not undergo meiosis, fertilization, and nuclear reorganization. A second kind of information exchange is required, certifying that the paired cells are of contrasting types, before the meiotic signal is given. This meiotic message is being studied by Miyake and his associates in an especially imaginative way. As with many ciliates (Chapter 10), Blepharisma may occasionally form symmetrical doublet cells capable of more or less indefinite persistance through cell division. By treating a population of doublets of one mating type with the gamone of the other type, homotypic agglutination of the doublets is induced. Because each of the doublets has two mating surfaces, linear arrays of doublets may develop (Fig. 4–4), and these may even close into completed rings. Meiosis does not occur in these homotypic assemblages. One may, however, attach to the end of the homotypic array a single cell of the other mating type, and this heterotypic association generates the meiotic message. The passage of this message through the array may be followed by the sequential occurrence of meiosis through the homotypic chain. Indeed one may, after a period of time, remove the singlet, thus blocking the formation of additional meiotic message, and trace the passage of the signal, measuring its rate of movement and its limits by diffusion.

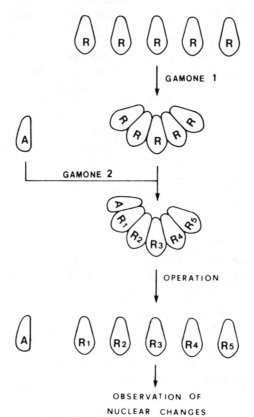

GAMONE 1

GAMONE 2

OPERATION

OBSERVATION OF
NUCLEAR CHANGES

Fig. 4-4. Experimental design for the study of cell-to-cell propagation of nuclear activation in *Blepharisma japonicum*. Red doublet cells of mating type II (R) in the presence of gamone 1 form homotypic chains, unable to begin meiosis. A singlet albino cell of type I (A) is activated by gamone 2 and is allowed to attach to the end of a homotypic chain. At various times after attachment samples of the chains are removed and the components are separated in sequence (A, R_1, $R_2 \ldots R_N$) and observed for evidence of nuclear reorganization. Reprinted, permission from Miyake, A. 1978. Cell communication, cell union and initiation of meiosis in ciliate conjugation. *Curr. Topics Dev. Biol. Fertilization* 37–82. Fig. 17, p. 70. Copyright by Academic Press Inc. (London) Ltd.

Homotypic pairing also occurs in some of the hypotrich ciliates, following a complex behavioral interaction (Fig. 4-5). In the genetic species of *Euplotes patella* studied by Kimball and Powers, as in *B. japonicum,* homotypic pairing is induced by cellfree fluids from cultures of other mating types. Unlike the

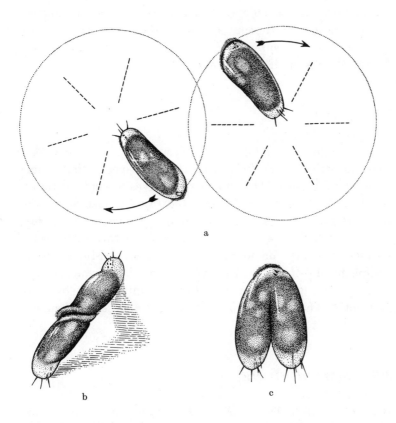

Fig. 4-5. Preconjugal interactions in *Stylonychia mytilus.* (*a*) Rotary motion of prepartners while attached to substrate. (*b*). Tete à tete attachment of partners. (*c*) Final side by side position of conjugants. Reprinted, by permission, from Grell, K. G. 1973. *Protozoology,* Springer-Verlag. Fig. 280, p. 326.

situation in Blepharisma, however, the homotypic pairing in Euplotes continues into meiosis and nuclear reorganization. The gamones in Euplotes have not been characterized, even though considerable work was done on the genetics of the multiple mating type system.

We cannot survey all the information available on cellular interactions in the ciliates. Ciliates are enormously diversified organisms with different communications requirements; these have been met by various signaling methods that may or may not be homologous in different groups.

SUMMARY

Many ciliates are grossly "underclassified." A "named species" may consist of several fully isolated sibling species that cannot be easily distinguished. The ciliates themselves recognize differences, however, and gene flow among the cryptic species is rare or nonexistent. The regulation of mating is achieved by surface properties that allow the organisms to assort themselves within a species into two or more "mating types." Mating between organisms of different mating type is controlled by an exchange of signals, acting through membrane contacts or by soluble molecules. The signalling devices may not be equivalent in different species.

RECOMMENDED READING

Bomford, R. 1966. The syngens of *Paramecium bursaria*: new mating types and intersyngenic mating reactions. *J. Protozool.*, **13,** 497–501.

Bruns, P. J. and R. F. Palestine. 1975. Costimulation in *Tetrahymena pyriformis*: a developmental interaction between specially prepared cells. *Dev. Biol.*, **42,** 75–83.

Esposito, F., N. Ricci, and R. Nobili. 1976. Mating-type-specific soluble factors (gamones) in cell interaction of conjugation in the ciliate *Oxytricha bifaria. J. Exp. Zool.*, **197**, 275–282.

Gruchy, D. G. 1955. The breeding system and distribution of *Tetrahymena pyriformis. J. Protozool.*, **2**, 178–185.

Jennings, H. S. 1939. Genetics of *Paramecium bursaria* I. Mating types and groups, their interrelations and distributions; mating behavior and self fertility. *Genetics*, **24**, 202–233.

Kimball, R. F. 1943. Mating types in the ciliate protozoa. *Q. Rev. Biol.*, **18**, 30–45.

Luporini, P. and F. Dini. 1975. Relationships between cell cycle and conjugation in 3 hypotrichs. *J. Protozool.*, **22**, 541–544.

Miyake, A. 1968. Induction of conjugation by chemical agents in Paramecium. *J. Exp. Zool.*, **167**, 359–380.

Miyake, A. 1974. Cell interaction in conjugation of ciliates. *Curr. Topics Microbiol. Immunol.*, **64**, 49–77.

Miyake, A. 1978. Cell communication, cell union and initiation of meiosis in ciliate conjugation. *Curr. Tropics Dev. Biol.*, **12**, 37–82.

Nanney, D. L. 1977. Cell–cell interactions in ciliated protozoa: evolutionary and genetic constraints. In *Microbial Interactions* (J. Reissig, Ed.), Chapman and Hall, pp. 351–397.

Nanney, D. L. and J. W. McCoy. 1976. Characterization of the species of the *Tetrahymena pyriformis* complex. *Trans. Am. Microsc. Soc.*, **95**, 664–682.

Siegel, R. W. and L. L. Larison. 1960. The genetic control of mating types in *Paramecium bursaria. Proc. Natl. Acad. Sci. U.S.*, **46**, 344–349.

Sonneborn, T. M. 1939. *Paramecium aurelia*: mating types and groups; lethal interactions; determination and inheritance. *Am. Nat.*, **73**, 390–413.

Sonneborn, T. M. 1975. The *Paramecium aurelia* complex of 14 sibling species. *Trans. Am. Microsc. Soc.*, **94**, 155–178.

Wellnitz, W. R. and P. J. Bruns. 1979. The pre-pairing events in *Tetrahymena thermophila. Exp. Cell Res.*, **119**, 175–180.

Wolfe, J. 1976. G_1 arrest and the division/conjugation decision in Tetrahymena. *Dev. Biol.*, **54**, 116–126.

Wolfe, J. and G. W. Grimes. 1979. Tip transformation in Tetrahymena: a morphogenetic response to interactions between mating types. *J. Protozool.*, **26**, 82–89.

Evolutionary Relations
and
Evolutionary Distances

5

BIOLOGISTS MUST KEEP IN MIND THE HISTORIES OF their organisms, because explanations of biological phenomena based solely on contemporary considerations are almost certain to be flawed. Organisms are the way they are in part because of their itineraries. Evolutionary opportunism and evolutionary inertia account for more of the details of biological structures and functions—at all levels of organization—than we usually admit.

For this reason we need to examine briefly the origins of the ciliates and their place in the evolutionary scheme of things, not so much because our certain knowledge of these matters illuminates, but because the recognition of our ignorance may forestall error and provoke new studies.

A. Evolutionary Relationships of the Ciliates

The largest obvious discontinuity among extant life forms is that which separates the eukaryotes from the simpler protists. The discontinuity involves several organellar distinctions: the possession of mitochondria and a phagocytic membrane, the capacity to produce the 9 + 2 cilium—flagellum and its derivatives, the presence of a membrane surrounded nucleus and a nucleosomic organization of the chromosomes, and the capacity for mitosis and meiosis. According to all these criteria the ciliates lie unambiguously on the side of the eukaryotes. The problem of their placement is in determining their relationships to the other eukaryotes—the fungi and algae and the multicellular plants and animals.

One consideration bearing on the evolutionary events is the time at which the eukaryotic life style was invented. The course of evolution involves an almost unimaginable time scale, and the fixed points of reference have a disconcerting tendency to become unfixed as studies continue. The estimated age of the earth does seem to have stabilized, however, at about 4.5–5 billion years (Fig. 5–1). The time at which evolving life forms appeared on the primitive earth has been gradually pushed backward in recent years. The oldest known rocks are only about 4 billion years old, and these contain no certainly identified fossils. However, the earliest living creatures were probably simple and delicate, and difficult to preserve. Their apparent absence from the oldest rocks is not decisive evidence for their nonexistence when the rocks were formed. Indeed, many geochemists now believe that even the oldest rocks bear evidence of the chemical activities of primitive life forms.

Perhaps then we can accept provisionally the time of 4 billion years ago (BYA) as the time of one of the most fundamental of biological inventions—that necessary for the linear processing of chemical information. This invention was a complex evolutionary event. It involved the choice of sugars and nitrogenous bases for nucleic acids, the choice of amino acids for polypeptides, the development of a codon dictionary relating nucleic acid components to protein components, and the construction of a bipartite ribosome to mediate between the major macromolecular languages. Since all modern life forms have essentially the same fundamental mechanisms for the processing of macromolecular specificities, they all almost certainly trace directly to the same time, the same place, and the same biochemical matrix.

The first primitive information processing machine, designated the Progenote by Woese, seems to have given rise to three major lineages (Urkingdoms) still represented in modern life forms, the familiar Prokaryotes, the Eukaryotes, and a group of previously unsuspected distinction—the Archaebacteria. This trichotomy, based on a comparative study of the nucleotide compositions of the ribosomal RNA, has two possibly surprising elements. The first, of course, is the existence of the "Third Kingdom," relict organisms driven today into bogs, ocean depths, and salt brines, where they maintain existence in environments related to primitive earth conditions or at least environments in which their special adaptions enable them to keep at bay the species of other

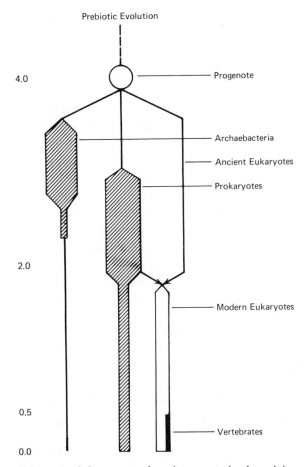

Origin of Earth

Prebiotic Evolution

4.0 — Progenote

Archaebacteria

Ancient Eukaryotes

Prokaryotes

2.0

Modern Eukaryotes

0.5

Vertebrates

0.0

Fig. 5-1. Diagram of the proposed major events in the origin and evolution of life on earth, beginning with the origin of the earth 4.5–5.0 BYA. Prebiotic and protobiotic evolution led by about 4.0 BYA to the Progenote, the organismic inventor of the present system of macromolecular information management. Derived from this common source are the three Urkingdoms now recognized—the Archaebacteria, the Prokaryotes, and the Ancient Eukaryotes. Modern Eukaryotes are proposed to be derived through one or more symbiotic interactions of some unspecified nature between one or more prokaryotes and one of the ancient eukaryotes. This junction is placed at about 2.0 BYA. For temporal scale the origin of the vertebrates at about 0.5 BYO is indicated. The ciliates are believed to have arisen immediately after the origin of modern eukaryotes.

69

lineages. The second possibly unexpected feature is the antiquity of the separation of the eukaryotic genome from that of the other lineages. A common view until recently was that the eukaryotes were derived from one or more prokaryotes at a relatively recent time. This interpretation is denied by the ribosomal sequence data.

Eukaryotes, of course, possess at least two information processing systems (three in the case of photosynthetic plants), and the reader should be aware that we have thus far referred only to the cytoplasmic ribosomal components and not to the mitochondrial (or plastid) ribosomes. In agreement with other evidences of affinity (size and antibiotic sensitivity), the mitochondrial ribosomes, at least those of the higher plants, have unmistakable similarities to those of prokaryotes. One may argue that the sequence similarities between mitochondria and prokaryotes are the consequence of convergent evolution, but the simplest interpretation is that the modern eukaryotes have a dual history; their two information processing systems trace back to the Progenote by different paths.

If this interpretation is accepted, it brings us back again to the origin of the "modern" eukaryotes, defined now as the time of junction between the separate evolutionary lineages. One central issue is whether all eukaryotes are derived from a single junction or whether several different unions occurred at different times and between different components. Another question is when the junction(s) occurred.

Once again the fossil record provides only limited information. The first eukaryotes were certainly protists, and they may or may not have formed hard parts capable of being preserved. The techniques for evaluating microscopic fossil structures and distinguishing them from nonbiological artifacts are gradually pushing back the time of origin of eukaryotes. Unmistakable protists, interpreted as bacteria and prokaryotic algae, are described from rocks as old as 3.4 billion years. Generally accepted eukaryotic protists are certainly seen in rocks of 1.0–1.4 BYA, and the Gunflint cherts from Lake Superior, dated at 1.9 BYA, contain forms interpreted by some as being similar to modern protozoa. Some students believe that yeastlike structures can be identified in rocks over 2 BY of age. Present technology thus enables us to say with reasonable certainty that some eukaryotes were present by about 1.5 BYA, and possibly as early as 2.0 BYA, but the record is inadequate to provide as yet an indication of whether the various eukaryotic forms arose quickly or over a long period of time. Just to keep things in perspective we should recall that this time interval we are considering, the 500 million years between 1.5 and 2.0 BYA, is equivalent in length to the entire evolutionary history of the vertebrates; the earliest known fishes occurred in the Ordovician period, about 450 million years ago.

The absence of an adequate fossil record should not discourage investigation of these questions. We have already mentioned plausible evolutionary reconstructions for an even more remote time. What is needed for a resolution of the early phylogeny of the eukaryotes is the judicious application of appropriate molecular chronometers. Many molecules change steadily over long periods of evolutionary time, some fairly rapidly, others at intermediate rates, and others, such as ribosomal RNAs, only very slowly indeed. Recent phylogenetic events may be reconstructed with fast chronometers, in which enough changes occur over the times investigated to be statistically meaningful. Fast chronometers eventually lead to only random similarities among compared structures, however, and have to be replaced by slower clocks that still permit relationships to be detected.

If chronometric molecules maintain their characteristic rates of change, we need only to choose the appropriate molecules to identify the approximate times at which homologous molecules in different organisms had a common ancestor. Unfortunately, we cannot always be sure that the rates of substitution of bases or amino acids are constant. Probably they are reasonably constant over some periods of time, but the rates may be very different in times of biological revolution, such as the time immediately following the invention of the ribosome or the time at which a primitive eukaryote began exploring its new capabilities. Such considerations should not discourage the use of molecules in reconstructing phylogenic histories; they only suggest the need for caution in interpretation and for the prudent use of several different molecular probes. Unfortunately, we do not as yet possess enough comparative information on the anatomy of molecules to place the ciliates with certainty. Two recent and ongoing studies of the kind that should lead to a definitive conclusion are discussed below.

The cytochrome c molecule is an ancient and conservative component of the respiratory system and is distributed almost universally among modern organisms. Its rate of change over recent geological epochs that can be monitored by fossil records appears to be fairly steady; substitutions of amino acids occur about three times per 100 million years, giving a 1% change of amino acid sequences in about 20 million years. If we assume that prokaryote and eukaryote cytochromes c have a common origin, the separation of prokaryote and eukaryote lineages can be estimated at about 1.9 BYA. This observation can provide a fix for the time of junction of the primitive eukaryote and its hypothetical mitochondrial symbiont. Geochemists also note evidence for a sharp increase in atmospheric O_2 about 2 BYA and believe that oxides signal the increased activity of the oxygen generating eukaryotes. Interpretation is clouded, however, by the still unresolved questions of the symbiotic origin of the mitochondria and of the evolutionary origin of cytochrome c. The mitochondrion houses the respiratory system and contains the cytochrome c, but the gene specifying cytochrome c is usually a nuclear gene. Some investigators believe that the prokaryotic cytochrome gene was captured with the mitochondrial genome and may have been secondarily transferred to the eukaryote nucleus. This interpretation allows a single "invention" of cytochrome c, in the prokaryote lineage after the Progenote had given rise to the primordial kingdoms, and it also provides a meaningful date for the prokaryote–eukaryote junction.

Somewhat dissociated from these speculative matters is the question of the relationships between the ciliates and the other eukaryotes. One interesting suggestion concerning the origin of the higher animals uses the ciliates as an intermediate. The increase in the size and compoundness by the ciliates has been proposed as the first step toward multicellularity. The compartmentalization, which permitted further specialization of parts and increase in size, was considered a second stage. Although perhaps plausible, this interpretation is not supported by the limited comparative biochemistry now available. Only one ciliate cytochrome c has been sequenced thus far, by Tarr, and it does not appear to be similar to that of the higher animals. Indeed, the Tetrahymena cytochrome c is the most distinctive of the eukaryotic molecules examined. If we assume that a prokaryotic cytochrome c gene was transferred to the eukaryotes, by whatever means, and if that transfer occurred only once, then the ciliates were the first group separated from the main eukaryotic trunk, prior to the Physarum–Crithidia–Euglena junction, and even prior to the higher plant junction on the way to the higher animals.

This preliminary indication of ciliate biochemical distinction is now strengthened by observations on yet another conservative molecule—one of the histones. Among the primary eukaryotic inventions is the nucleosome, which permits the management of up to 1000 times as much DNA as does the prokaryotic genomic device. The DNA in eukaryotes is associated with five kinds of histone molecules in equimolar amounts: H1, H2A, H2B, H3, and H4. The last four make up the octomeric core of the nucleosome. Of all these histones, the H4 is the most conservative, and indeed it has been labeled the most highly conserved protein known. Comparisons of H4 histone from calf and pea show only two amino acid replacements in the 100+ amino acids, and both of these are replacements by similar amino acids. The sea urchin

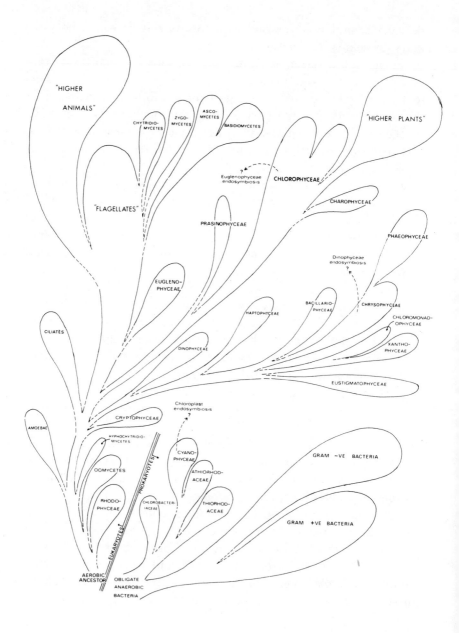

74

shows only one difference in comparison with calf. Generally the H4 histones from different eukaryotes can be freely substituted between species in *in vitro* reconstitution and affinity studies.

Yet, the Tetrahymena H4 molecule, in comparison with the calf molecule, is very different. In the first 66 residues examined, Glover and Gorovsky found 1 deletion, 1 insertion, and 13 replacements, of which only 4 were considered "conservative" replacements by similar amino acids. Two of the substitutions would require at least two base changes in the DNA. If, as suggested on the basis of other histone work, one substitution had occurred every 400 million years, the common ancestor of mammals and ciliates must be sought at 6 BYA, an unlikely projection considering other cosmic events at the time. Although histone 4 is not as conservative, and perhaps not as chronometric, as was previously believed, these preliminary studies again support the notion of a very distant relationship of the ciliates to the higher animals, and also the suggestion that the ciliates may have been one of the first of the eukaryotic groups to begin to explore their enlarged capabilities (Fig. 5–2).

The fragmentary data now at hand thus permit the proposal that the ciliates were among the first of the eukaryotes and that their explosive proliferation may have established the basic ciliate forms at a very ancient time, nearly 2 BYA. This interpretation is at least consistent with some puzzling observations concerning molecular diversity among the ciliates.

←───────────────────────────────────

Fig. 5–2. A proposed phylogeny of the protists based on available biochemical evidence of several sorts. In this proposed phylogeny the distance between groups, or the distance from the common ancestor, does not indicate the degree of divergence or a "time factor" in the evolution over geological time. Several such phylogenies have been proposed with differing details depending on which of the fragmentary evidences is emphasized. Reprinted with permission from Ragan, M. A. and D. J. Chapman. 1978. *A Biochemical Phylogeny of Protists,* Academic Press, Fig. 23, p. 221. Copyright by Academic Press Inc (London) Ltd.

B. Evolutionary Distances Among Ciliate Species

The occurrence of isolated breeding systems of similar ciliates is no major surprise. Many organisms have clusters of related species separated at various "distances" from other clusters of related species. The phenotypic similarities within the clusters suggest that these are *sibling species* of recent common origin and that they are similar because they have not had sufficient time to move apart.

The interpretation is challenged, however, as soon as we look at measures of molecular or genetic distance instead of phenotypic or phenetic distance. Strains of the *T. pyriformis* complex, incapable of being separated by size, cortical geometry, nutritional requirements or chromosome numbers, may vary dramatically in their molecular composition. Their DNA composition, for example, may vary from 25% guanine + cytosine to 33% guanine + cytosine. This 8% difference in DNA composition within the species complex is about twice the total variation observed among all the vertebrates. This indication of DNA differences is supported by DNA hybridization studies. Unique DNA sequences (to avoid possibly evolutionarily labile repetitive sequences) have been cross-annealed between strains and have been found to give as little as 10–20% of the duplexes formed with homologous mixtures.

Similar indications of molecular distances among ciliate species have been obtained with RNA–DNA hybridization studies and with studies of the electrophoretic mobilities of enzyme activities (isozymes). In some comparisons between strains of Tetrahymena no common electrophoretic mobilities were found when 10 enzyme activities were assayed. Studies of antigens and of structural proteins are consistent in indicating very different molecular compositions and, hence, in suggesting large genetic distances between phenotypically very similar organisms.

The molecular distances require that the strains must have had a very distant common origin, if the rates of diversification in ciliates are comparable to those in other organisms that have been studied. Even if we assume that the ciliates are unusually plastic in their genetic makeup, and few other indications of such plasticity are available, then we must explain why phenotypic plasticity is not associated with the molecular plasticity. The simplest hypothesis at the moment is that ciliates are special only in their evolutionary age and that the rates of molecular diversification are similar to those in other organisms.

This interpretation brings into focus two different connotations of the term evolutionary distance and suggests that genetic and phenetic distances, though highly correlated within a group of close relatives over a short evolutionary time, may be essentially disconnected when comparisons are made among groups of very different evolutionary age. The reason similar Tetrahymena species are alike is not that they have had insufficient time to diversify. They are alike because they are constrained within narrow limits to a "design," of scale, of form, of nutrition, of kar-yotype, arrived at long ago and little affected by subsequent biotic and geologic changes around them. Their constraint within these limits does not prevent molecular substitutions, which are permit-ted to continue so long as the basic design elements are maintained and may be required for environmental "tracking."

The protozoa offer an interesting and provocative contrast to groups of organisms of recent origin. The vertebrates have undergone an enormous phenetic "explosion" in a relatively short time, but the molecular changes underlying the phenotypic diversity are relatively slight. We look for molecular differences in the vertebrates in the hope of discovering the bases for differences in form and function, but in Tetrahymena we must look for molecular similarities—to identify the basis for the unchanging design.

C. Code Limit Organisms

Before leaving the subject of molecular composition, a brief diversion should be made concerning the low frequency of guanine + cytosine in the ciliate DNA. Many organisms have about equal frequencies of guanine, cytosine, adenine, and thymine in their DNA. This near equality of the bases led to the unfortunate "tetranucleotide hypothesis" of DNA structure that temporarily obscured the genetic role of DNA. The departures from equality, observed when larger arrays of organisms were examined, permitted the realization that compositional variation was possible, but that the variation was constrained, that is A = T and G = C. This rule was an important component in establishing the double-helix structure of DNA.

Why do most organisms have DNA compositions of approximately equal amounts of the four bases? Perhaps the best reason available for a "balanced composition" is that it provides the largest flexibility for the use of codons in the DNA. If an organism has a low frequency of C + G, then codons with C and G will be relatively rare. Amino acids coded by triplets of CCC, GGG, CGC, and so on will have to be used infrequently, and amino acids coded by triplets of AAA, TTT, TAT, and so on will become more common. An organism with a high proportion of G + C must modify its protein composition in the other direction and is unable to use certain other codons freely because of their rarity. In contrast, an organism with near equality of base composition may use the entire codon dictionary. The limits of base composition on codon assignments are relieved somewhat by the degeneracy of the code, but clearly the base composition of the DNA affects the amino acid composition of the catalytic and structural proteins.

Why then do some ciliates have such eccentric base ratios? Why have they become "code-limit organisms"? The answer is not known, but the kinds of selective pressure that might be significant can be mentioned. Growth in an environment notably deficient in the precursors of cytosine and/or guanine might be expected to drive the base ratio downward. Similarly, an environmental limitation in the amino acids coded by high GC codons, or in precursors of these amino acids, could secondarily force the base ratio downward. Finally, the DNA composition itself, or the amino acid composition, might be a phenotype upon which natural selection could obtain leverage. Present information about the nutritional requirements and the food habits of the ciliates bear on these possibilities, but do not permit a resolution.

Finally, we should point out that all groups of ciliates may not be as molecularly diversified as are the Tetrahymena. The *P. aurelia* group, for example, seems much less heterogeneous in molecular composition. Perhaps a series of ciliate species clusters of different evolutionary age can eventually be developed and then used to answer some interesting questions.

SUMMARY

Because the ciliates are not well represented in the fossil record, an understanding of their early history and their organismic relationships must rely heavily on the evidence of comparative anatomy, and particularly the comparative anatomy of molecules. Because chronometric molecules are only now being calibrated, and few have been examined in the ciliates, very few interpretations concerning relationships among the ciliates are firmly founded. Even the relationship of the ciliates to other eukaryotes is problematical.

The ciliated protozoa may, however, have emerged as one of the earliest eukaryote experiments, possibly as early as 2 BYA. Most of the modern orders may have emerged during that early radiative period, possibly over a very short period of time. If the interval of radiation was short, relative to the elapsed time since the radiation, molecular chronometry may not give decisive answers about intraciliate affinities. The chronometric scrambling that has occurred since the major groups emerged may obscure the significant early events. In any case, the molecular distances between some sibling species suggest that even these separations may have occurred at a very remote time.

RECOMMENDED READING

Allen, S. L. and C. I. Li. 1974. Nucleotide sequence divergence among DNA fractions of different syngens of *Tetrahymena pyriformis*. *Biochem. Genet.*, **12**, 213–233.

Borden, D., E. T. Miller, G. S. Whitt, and D. L. Nanney. 1977. Electrophoretic analysis of evolutionary relationships in Tetrahymena. *Evolution*, **31**, 91–102.

Glover, C. V. C. and M. A. Gorovsky. 1979. Amino-acid sequence of Tetrahymena histone H4 differs from that of higher eukaryotes. *Proc. Natl. Acad. Sci. U.S.*, **76**, 585–589.

Goldbach, R. W., A. C. Arnberg, E. F. J. Van Bruggen, J. Defize, and P. Borst. 1977. The structure of *Tetrahymena pyriformis* mitochondrial DNA. I Strain differences and occurrence of inverted repetitions. *Biochem. Biophys. Acta*, **477**, 37–50.

Hutner, S. H. and J. O. Corliss. 1976. Search for clues to the evolutionary meaning of ciliate phylogeny. *J. Protozool.*, **23**, 48–56.

Johmann, C. A. and M. A. Gorovsky. 1976. An electrophoretic comparison of the histones of various strains of *Tetrahymena pyriformis*. *Arch. Bioch. Biophys.*, **175**, 694–699.

Nanney, D. L., D. Nyberg, S. S. Chen, and E. B. Meyer. 1980. Cytogeometric constraints in Tetrahymena evolution: contractile vacuole positions in 19 species of the *T. pyriformis* complex. *Am. Nat.*, **115**, (in press).

Nobili, R., P. Luporini, and F. Dini. 1978. Breeding systems, species relationships and evolutionary trends in some marine species of Euplotidae (Hypotrichida Ciliata). From *Marine Organisms* (B. Battaglia, J. Beardmore, ed.), Plenum, 591–616.

Ragan, M. A. and D. J. Chapman. 1978. *A Biochemical Phylogeny of Protists*, Academic.

Tait, A. 1978. Species identification in protozoa: Glucosephosphate isomerase variation in the *Paramecium aurelia* group. *Biochem. Genet.*, **16**, 945–955.

Vaudaux, P. E., N. E. Williams, J. Frankel, and C. Vaudaux. 1977. Interstrain variability of structural proteins in Tetrahymena. *J. Protozool.*, **24**, 453–458.

Woese, C. R. and G. E. Fox. 1977. Phylogenetic structure of the prokaryotic domain: The primary kingdoms. *Proc. Natl. Acad. Sci. U.S.*, **74**, 5088–5090.

Mating Tactics
and
Ecogenetic Strategies

6

A. Balancing Stability and Variety

The sibling species of a ciliate species complex are very similar in their general form, that is, their size, appearance, nutrition, and habitat. Yet their molecular diversification indicates that they may have been separated for enormous lengths of time (Chapter 5). Even though two or more different species may be collected from the same stream or pond, indeed from a single small water sample, little is known about their interactions in nature.

Few studies of ciliates in their natural environments are available, and we know little about environmental factors affecting the success of particular species. Especially rare are ecological studies that discriminate among the species of a complex. Laboratory competition among strains of different species complexes has been studied occasionally, but these experiments provide little understanding of the long term coexistence of sibling species.

In the absence of information concerning relevant environmental variables, attention has been focused on variable strategies for the utilization of a possibly common environment. This approach requires, perhaps, a brief rationalization. All organisms must achieve an appropriate balance between stability and flexibility. An organism that is too stable is unable to change as conditions inevitably change—whether on a daily, seasonal, or geological time scale—and is replaced by organisms that can adjust. Variability also has its hazards, however, and its cost. Genetic variation, for example, may come by way of mutation. Mutations are undirected molecular alterations of the genetic apparatus whose consequences for the organism cannot be predicted in advance. Understandably, most mutations, though they provide variety, provide a useless kind of variety and lead to the demise of the organism carrying them or of a descendant. Natural selection then tends to promote a mutation rate that is sufficiently high for the organism to "track the environment," but a rate that is sufficiently low that progeny inviability (genetic load) is kept at tolerable levels.

The ecogenetic strategies that distinguish ciliate species have to do with the management of this stability–variety dilemma. Above, mutation is mentioned as a component to be adjusted, but mutations represent only one means of achieving genetic variety. Genetic variability can also be generated by recombination of genetic components (chromosomes, cistrons, base pairs) of different origin. Like mutation, recombination also affects the phenotypes of the organism—and in a somewhat unpredictable fashion. Yet, generally, recombination tends to be a more conservative means of generating genetic diversity, because the components to be recombined (mutations ultimately) have already passed through a selective filter. Recombinants are more likely to be viable, but also are less likely to yield a dramatically different phenotype.

Since mutation and recombination provide somewhat different kinds of genetic variability, both are usually retained in the genetic economy, but achieving the right proportions requires a balancing act. What is the proper ratio of mutation to recombination, and how may it be regulated? Here we have to consider briefly a significant difference between organisms that have an extended haploid phase in their life cycle and those that do not. When a mutation occurs in a haploid organism it is immediately exposed to environmental evaluation. If the variant is harmful the organism dies, but if it improves the organism's fitness it has the opportunity to be multiplied in the population. Since most mutant genes are recessive, however, a mutation in a diploid organism is not immediately exposed to the full effects of natural selection; it is usually "covered" or tempered by an allele with acceptable properties. If a haploid and a diploid population have the same frequency of a particular mutant gene, they have very different frequencies of the mutant phenotype. If for example the mutant gene has a frequency q, then the mutant phenotype in the haploid population is q, but in the diploid population it is q^2; when the mutant gene is rare, the frequency of mutant individuals (homozygotes) is very low indeed.

A haploid organism thus has available for immediate use any mutations that have occurred in the recent past. It can adjust to environmental changes (antibiotics or unusual substrates, for example) by utilizing the mutational variety at hand. A diploid organism, in contrast, though it may have the same frequency of mutant genes, has most of those genes stored in an unexpressed and hence useless condition. Indeed the diploid organism usually has far more genetic variability hidden in its genome, because it is accumulated without being expressed; but that variability cannot be readily deployed to meet sudden environmental changes. The ability of a diploid to meet environmental challenges by mutational changes can only approach as a limit the ability of the haploid, and then only after a long period of mutation accumulation, when nearly every genetic locus is heterozygous. For such reasons, *"the haploid strategy"* may be referred to as a distinctive genetic economy based strongly on mutation to supply the needed genetic variety.

One might then be inclined to characterize *"the diploid strategy"* as a genetic economy in which genetic variety is provided by recombination. This solution is too simple, however. In the first place, haploid organisms also utilize recombination, as is shown by the pervasiveness of sexual and parasexual systems in most bacteria, algae, protozoa, and fungi. Secondly, sexuality does not automatically provide the release of genetic variability that one might imagine. In fact, a randomly cross-breeding diploid organism has access to no more mutational variety than does an asexual diploid population. The frequency of homozygotes after a random sexual orgy is still q^2, the same frequency achieved by random mutation.

A diploid organism may, however, achieve the same level of mutational variety available to a haploid by modifying its breeding strategy. A diploid organism may, for example, undergo a kind of sexual reorganization called *autogamy,* in which gamete nuclei derived from the same meiotic product unite to form a zygote. All such zygotes are homozygous, and all mutations are expressed. A diploid organism with compulsory autogamy has adopted the haploid strategy and can readily adapt to environmental vicissitudes by means of mutational changes.

Few diploid organisms revert completely to a haploid economy, but many make an accommodation toward homozygosity. Any organism, for example, that mates preferentially with "close relatives" is more likely to produce homozygotes and expose mutations than is an organism that systematically mates with "strangers." By adjusting the breeding patterns, one may develop a continuous distribution of organisms ranging from strictly autogamous forms, through close inbreeders to preferential outbreeders. Inbreeding and outbreeding are not simple alternatives, but a sliding scale capable of fine adjustment.

We have spoken of the advantages of inbreeding in terms of the utilization of mutational variety, but we have not characterized the advantages of outbreeding beyond suggesting that they have something to do with recombination. A definitive discussion of this issue is beyond the scope of this book and the competence of this writer, but a few brief comments might be helpful. Although inbreeding enhances the probability of expressing a mutation, it has a low probability of producing significant new combinations of genetic elements. The number of permutations and combinations is determined by the numbers of different elements to be combined, and highly inbred forms have little diversity. Sexual unions of sister cells are nearly useless in terms of generating variety and are usually avoided except when homozygosity per se is advantageous.

As the genetic distance between mating organisms increases, the number of combinations of genetic elements in their progeny rises sharply. Many of these recombinants have little biological significance, and some may manifest incompatibilities, but in some cases positive interactional combinations are achieved, and these can be differentially transmitted. As the genetic distance between the parents continues to increase, however, the incompatibilities begin to predominate over the useful combinations. "Coadapted gene complexes" are broken up and poor fitness results. Selective forces may then accumulate to inhibit intercourse with strangers, and the species begins to disintegrate into subspecies. The precise level of inbreeding or outbreeding that is optimal for a species is affected by the diversity of the environment in which it lives, by the genetic diversity of the species as a whole, and by the usefulness of mutational variety as opposed to recombinational variety in meeting the special kinds of challenges to which the species is exposed.

The outbreeding genetic economy should not, however, be thought of simply as a strategy for using recombinational variety as a substitute for mutational diversity. Genetic diversification is, after all, only one way to adapt to a changing environment. A population exposed to a new substrate may adapt by transcribing a previously silent gene instead of mutating a preexisting gene. Indeed, for commonly encountered environmental variables, a physiological adjustment that permits most of a population to survive may be preferable to a genetic adjustment that saves only a rare mutant. The cost of carrying unused equipment, the cost of sensing the need for such equipment, and the cost of mobilizing that equipment, must be evaluated against the benefits available when the equipment is required. Thus it is necessary to contrast not only mutational variety and recombinational variety, but genetic variety and regulative variety as means of meeting environmental changes. Although outbreeding organisms appear to be designed to generate recombinational diversity, they seem also to be much more adept at meeting changed conditions head on, by changing their patterns of synthesis and the components of their organelles. Outbreeders tend to be less specialized in terms of environmental requirements and more able to cope directly with changing conditions.

B. Comparative Genetic Economies

The considerations outlined above derive largely from Sonneborn's attempt to rationalize the striking differences in the life histories of different ciliates. Species or species complexes that seem otherwise very similar may show dramatic differences, particularly in the way they manage their sexual affairs. To illustrate this diversity, let us consider two groups of species—those of the *P. aurelia* complex and those of the *P. bursaria* complex.

We note earlier that the *P. aurelia* species all have two mating types, but the *P. bursaria* species have four or eight. What is the significance of this difference? Sonneborn noted that in a population composed of two equally frequent mating types, a chance encounter between two strangers would involve different mating types in one half of the cases. In contrast, if the population contained four mating types in equal frequency, then three quarters of all chance meetings of strangers could lead to a mating; and in a system of eight mating types, seven eighths of chance encounters are potentially sexual. If a species "wants" to increase the chance of mating with a stranger, it can do so by increasing the number of mating types, and hence the number of mating options. According to this consideration, the *P. bursaria* species are more committed to outbreeding than are the *P. aurelia* species.

A similar kind of argument is applicable to differences in the duration of their immaturity. The species of *P. bursaria* have a period of sexual incapacity that immediately follows conjugation. For dozens of cell divisions, and months of time, the progeny of a cross may not be able to mate with anything. In contrast, the species of the *P. aurelia* complex have no immaturity, or an immaturity period lasting only a few days, up to a week or so. Sonneborn explained immaturity as a device for regulating who mates with whom. Time and numbers of cell divisions are measures of the distance a ciliate has moved from its point of origin. At the time of the fertilization that produces a new clone, it is surrounded by parental cells and by siblings, but as time passes it gradually moves away from its geographic and genetic origin and comes to interact with strangers. The longer the period of sexual immaturity, the more likely is a cell to be among strangers when it is prepared to mate. According to this criterion also, *P. bursaria* is more of an outbreeder than is *P. aurelia*.

Another initially perplexing difference between these species complexes was noted in the manner of mating type determination. The *P.·bursaria* species all show *synclonal uniformity*. That is, all the offspring of any particular pair (a *synclone*) have the same mating type. This result is the expected consequence of direct genetic control of mating type, for all the cells of a synclone have the same genotype (Chapter 7). Indeed, the ratios of mating types among the synclones produced in crosses can be explained by simple genetic determination. In contrast, the synclones of species of the *P. aurelia* complex usually consist of mixtures of mating types. Mating types are "determined" by developmental events that occur after fertilization, and in such a fashion that most synclones are assured of some cells of each type.

We consider the mechanisms of mating type determination later (Chapters 13 and 14). For present purpose we are interested primarily in the consequences. Synclones of *P. aurelia* species regularly contain two mating types, and they rapidly become mature. Hence, mating among cells of the closest genetic relationship (identity) is easily achieved. The synclones of *P. bursaria*, even when they become mature after a long delay, are unable to participate in this kind of close inbreeding. Again *P. bursaria* demonstrates its commitment to an outbreeding economy.

There is not space enough here to describe how all the life history differences among these two groups of species can be explained by their genetic economies. As expected, the *P. aurelia* species have generated a greater number of species, and species of more limited distribution, than the outbreeding *P. bursaria* complex. The *P. bursaria* species are long-life forms, as is necessary for organisms that have to await the discovery of a suitable stranger. Perhaps the most persuasive evidence for the explanation of the differences between the groups is the distribution of autogamy. All the species of the *P. aurelia* complex undergo autogamy if they do not find mating partners within a reasonable time. Autogamy has never been found in species of the *P. bursaria* complex.

While the Sonneborn thesis is compelling when viewed against these two very different species groups, our original problem was to explain the continued coexistence of sibling species in the *same* complex. To approach this question we must first point out that species groups are not uniform in their measures of outbreeding commitment, even though species within a complex are more alike than species in different groups. Species of the *P. aurelia* group differ in their length of life and immaturity periods; they differ in their mode of mating type determination and in the frequency with which they are observed to undergo autogamy or selfing. The range of variation—on an inbreeding–outbreeding scale—is narrower for the complex than for the entire genus, but it is nevertheless considerable. The species of the *P. aurelia* complex do not have exactly the same genetic economies, and therefore they do not exploit precisely the same habitat. The distinctive ways in which they manage their requirements for stability and flexibility assure that they inhabit different environments—even though they may encounter each other in the same space.

Before leaving the subject of adaptations associated with the breeding structure of ciliate species, brief mention should be made of "escape hatches." Under certain unusual conditions strains may break the laws ordinarily governing their behavior and engage in illicit acts. These exceptional behaviors are often rationalized as built-in ecotactics justified by survival considerations. One such escape hatch is referred to as "senescent selfing." In one of the species of *P. bursaria* (as well as in some Euplotes species), occasionally clones have been observed after several years of culture to undergo *selfing,* that is, to form pairs within a culture originally pure for mating type, contrary to the usually strict prohibition of mating within a synclone in this species. This behavior may be understood as adaptive in a colonizing species, because it allows recombination to diversify a monotypic clone. It is particularly justified in this instance because the species has a finite life cycle (Chapter 8); after a certain number of cell divisions, the strains gradually lose their ability to divide, or to survive conjugation even if they find a mate. Under these circumstances, the usual regulations are suspended and intraclonal mating may occur (though only in certain genotypes); incest is preferable to extinction.

A second and similar example can be drawn from another ciliate, *Tetrahymena thermophila*. This species has a fairly long period of sexual immaturity, lasting usually 60–80 cell divisions. When, however, a conjugating pair and its early synclone are exposed to adverse conditions, the cells often accelerate their maturation and mate within 12–15 cell divisions. Curiously enough, mutations at many different loci can also give an *early maturity* phenotype, and all these mutants are characterized by slow and erratic growth. The common interpretation of these mutants and of their induced phenocopies is that the cells can perceive any serious incompatibility between their genetic heritage and their environment and can respond by short circuiting the usual interval of abstinence. Whether the incompatibility is due to an abnormal environment or an unusual genotype, the cell can return quickly to the sexual lottery and expect to produce some progeny more capable of coping than is the parent.

The ciliates provide a rich source of comparative materials for the study of ecogenetic strategies. They are particularly useful because species so similar in so many respects are so diversified in their breeding systems. *Tetrahymena rostrata* is an obligate inbreeder, forming cysts and committing autogamy whenever the food supply is exhausted; it has never been seen to mate. *Tetrahymena americanis* is an outbreeder, usually requiring 100–200 cell generations before reaching maturity and acquiring the ability to mate—and then only with a stranger. *T. pyriformis (sensu stricto)* is amicronucleate and incapable of mating with anyone under any circumstances. All these breeding strategies are successful, but we are only beginning to understand their implications.

SUMMARY

Although their organismic design places significant limitations on the size and distribution of ciliates, they have exploited their environments through an enormous diversity of ecogenetic strategies. Particularly, some ciliates are inbreeders, rely strongly on mutational variety to meet environmental challenges, and have genetically specialized local populations. At the other extreme are outbreeding ciliates that rely on recombinational variety and physiological plasticity.

These different strategies are evidenced by many details of the life histories of the species—the number of mating types, the length of the period of immaturity following mating, the total life span, the mode of mating type determination, and the occurrence of autogamy. The strategies may scarcely be detected, however, in other major features of organismic design, that is, in their size, shape, or nutritional requirements.

RECOMMENDED READING

Corliss, J. O. 1965. L'autogamie et le sénescence du cilié hyménostome *Tetrahymena rostrata* (Kahl). *Ann. Biol.*, **4**, 49–69.

Hairston, N. G. 1958. Observations on the ecology of Paramecium with comments on the species problem. *Evolution*, **12**, 440–450.

Karakashian, S. J. 1966. Mating in populations of *Paramecium bursaria*. A mathematical model. *Genetics*, **53**, 145–156.

Nyberg, D. 1974. Breeding systems and resistance to environmental stress in ciliates. *Evolution*, **28**, 367–380.

Orias, E. and Rohlf, F. G. 1964. Population genetics of the mating type locus in *Tetrahymena pyriformis*, variety 8. *Evolution*, **18**, 620–629.

Power, H. W. 1976. On forces of selection in the evolution of mating types. *Am. Nat.*, **110**, 937–944.

Sonneborn, T. M. 1957. Breeding systems, reproductive methods and species problems in Protozoa. In *The Species Problem* (E. Mayr, Ed.), American Association for the Advancement of Science, pp. 155–324.

Cytogenetic Maneuvers

7

A. Conjugation

Nuclear behavior at conjugation in ciliates appears at first to be a formidable barrier to comprehension of the organisms. One has to be concerned with both micronuclei and macronuclei, mitosis and meiosis, nuclear growth, differentiation, and disintegration through a complex series of stages leading from the original parental nuclear constitution to the derived nuclear apparatus of the progeny. A part of the difficulty in understanding the sequence is that all the different nuclear events occur in the same cytoplasmic arena, but at different times. The progeny are not provided with a cytoplasmic shell separate from that of the parents; instead the progeny nuclei replace the parental nuclei and come to occupy the same cytoplasmic domain. Much attention is given these days to "reproductive strategies," with respect, for example, to such matters as the merits of producing a few large eggs or many small eggs. The ciliates appear to have chosen an extreme position in this particular debate. The parent organism contributes its entire substance to a single offspring and obligingly removes itself from the scene. Two parent cells enter into conjugation, and after an interval two progeny cells emerge. (In some instances two parent cells enter conjugation and only one progeny cell emerges.)

The events associated with this change of administration may be separated into two major periods: the *prefertilization events* and the *postfertilization events*. Neither set of events is particularly complicated once the basic "objectives" have been grasped, but they may appear complicated because they are sometimes compound. To clarify the essential features we start with the simpler ciliates, such as Tetrahymena, which have only one micronucleus and one macronucleus (Fig. 7–1). In such a form the prefertilization events consist of (1) meiosis in the diploid micronucleus to produce four haploid products, (2) mitosis of *one* of the haploid products to yield a *migratory pronucleus* and a *stationary pronucleus*. (These two kinds of pronuclei were sometimes, in an era less

97

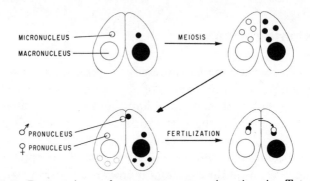

Fig. 7-1. Prezygotic nuclear events at conjugation in *Tetrahymena thermophila.* Each micronucleus undergoes meiosis to produce four haploid products. One haploid nucleus in each conjugant attaches to the mating surface and divides mitotically to form the migratory (or male) and stationary (or female) pronucleus. The other haploid nuclei (relics) move to the posterior end of the cell and disintegrate. Exchange of the migratory pronuclei leads to mutual fertilization. Reprinted with permission from Nanney, D. L. 1964. Macronuclear differentiation and subnuclear assortment in ciliates. In *The Role of Chromosomes in Development* (M. Locke, Ed.) Academic Press. pp. 253–273. Fig. 1, p. 254. Copyright by Academic Press Inc. (London) Ltd.

sensitive about sex roles, designated as male and female pronuclei), and (3) removal of the superfluous haploid nuclei (relics).

The ciliate micronucleus does not have a perfectly conventional structure or behavior even in vegetative divisions. We mention earlier that it is physiologically very nearly inert. Moreover, the nuclear membrane persists through nuclear divisions, and the spindle fibers are arrayed inside the nuclear envelope. The chromosomes usually do not condense into discrete and readily counted bodies, but remain as tangled threads difficult to discriminate. It is perhaps surprising that the micronuclei are so nearly normal once meiosis gets underway. Except for a still poorly understood stage of elongation (the *crescent* stage) during prophase I, meiosis appears to be reasonably conventional.

The paired homologues are readily seen and counted (at least in forms with few chromosomes such as Tetrahymena), and meiosis continues as expected. The haploid number of chromosomes is easily demonstrated at the end of the second meiotic division.

The simultaneous division of one haploid product and the destruction of the other three is managed by an interesting system of nucleo-cytoplasmic interactions. The haploid nuclei are apparently attracted to a cytoplasmic shelter, located in the region of contact between the mates. This privileged site is capable of protecting any nucleus reaching it, but only one nucleus ordinarily reaches it; the nuclei outside the region (relics) are subject to degradation and are eventually destroyed completely. The protected nucleus, on the other hand, is stimulated to divide to produce the genetically identical pronuclei, one remaining in the sheltered area, which is also the point of nuclear transfer, and the other moving into another special region preparatory to receiving the migratory pronucleus.

Just before fertilization then, the migratory pronuclei from the two mates occupy positions near each other in the contact zone, but slightly displaced to each cell's right. Fertilization occurs with the simultaneous passage of the migratory pronuclei through the contact membranes to fuse with the waiting stationary nuclei. The fusion yields an identical *synkaryon* or *fertilization nucleus* in each cell. The mates may then separate, for their significant interaction is complete.

At the time of fertilization, therefore, each mating cell has been restored quantitatively to its original condition prior to conjugation; each cell has one micronucleus and one macronucleus. The significant difference is that the genetic constitution of the micronucleus may have been changed and may, therefore, have become different from that of the macronucleus in the same cell.

At this time the offspring is represented only in essence, in the quantitatively insignificant and physiologically inactive micronucleus. But this nucleus is programmed to take over the parental framework. It does this by the following maneuvers (Fig. 7-2): (1) The fertilization nucleus divides (usually twice), mitotically. (2) Some (usually two) of the nuclear products are induced to become compound; these are the *macronuclear primordia* or *anlagen*. (3) The old macronucleus is destroyed.

This sequence of events again involves a series of nucleo–cytoplasmic interactions. The second postzygotic division places identical daughter nuclei at opposite ends of the cell. The cytoplasmic conditions at or near the anterior end (in Tetrahymena; in Paramecium, the posterior end) induce nuclei entering the region at this time to begin DNA synthesis. Nuclei at or near the posterior

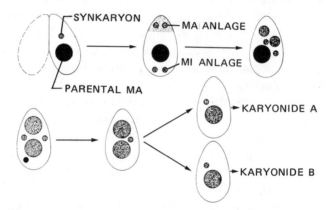

Fig. 7-2. Postzygotic nuclear events at conjugation in *Tetrahymena thermophila.* Each synkaryon divides twice mitotically, placing two diploid nuclei at each end of the exconjugant. The anterior nuclei continue replicating their DNA and are the macronuclear primordia. The old macronucleus and one of the micronuclei are destroyed. At the first cell division after conjugation the micronucleus divides, but the new macronuclei are assorted to the two daughter cells. These daughters are the origin of clones sharing a common macronuclear origin—karyonides. Reprinted, by permission, from Nanney, D. L. 1964. Macronuclear differentiation and subnuclear assortment in ciliates. In *The Role of Chromosomes in Development* (M. Locke, Ed.), Academic Press, pp. 253–273. Fig. 1, p. 254. Copyright by Academic Press, Inc. (London) Ltd.

end do not receive this signal and do not start a process of compounding, but remain micronuclei until the next conjugation. The new macronuclei apparently transmit a signal to the old macronucleus at the time that they begin to enlarge, stimulating it to begin a process of autolysis. Whenever macronuclear anlagen fail to develop properly, the old macronucleus remains intact and functional. In the normal course of events, however, the new macronucleus grows, the old macronucleus is destroyed, and the micronuclei persist. Actually, in Tetrahymena only one of the two micronuclei persists, and the other is destroyed, like the old macronucleus and the haploid relics; the mechanism whereby one presumptive micronucleus is chosen to survive has not been discovered.

These events complete the process of *nuclear reorganization,* except for an assortment phase. Each exconjugant cell has two macronuclear anlagen, while a normal vegetative cell has only a single macronucleus. The vegetative state is achieved by a cell division at which the new macronuclei are separated but do not divide. The micronucleus does divide, so that each of the two daughter cells then has one micronucleus and one macronucleus. These first cells with standard nuclear equipment have a special place in ciliate studies and give rise to specially designated clones, called *karyonides.* Since each conjugating pair produces two exconjugants, and each exconjugant produces two karyonides, each pair produces four karyonides, that is, clones bearing descendants of the four new macronuclei produced at nuclear reorganization. Karyonides are important because developmental events may bring about different characteristics of macronuclei with the same initial genotypes. We return to a consideration of "karyonidal determination" at a later time (Chapter 13). We should emphasize, however, that the four karyonides from the same conjugating pair are genetically equivalent, at least at first.

The nuclear events in Tetrahymena conjugation appear to be less complicated than those in Paramecium (Fig. 7–3), but a care-

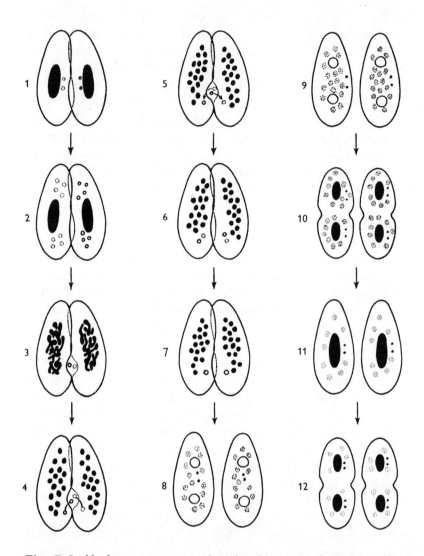

Fig. 7-3. Nuclear events at conjugation in one of the species of the *Paramecium aurelia* complex. (*1*) Paired cells with the normal vegetative complement of one macronucleus and two micronuclei. (*2*) After meiosis, each cell contains eight haploid nuclei. (*3*) One of the eight nuclei in each

ful examination of the diagrams shows the essential similarity. *P. aurelia* begins with two micronuclei instead of Tetrahymena's one, and the macronucleus is dismantled and discarded less directly. The genetic considerations are identical.

The genetic consequences of conjugation may be seen (Fig. 7–4) by examining a cross between two cells homozygous for different alleles at the same locus. The homozygous A/A cell can produce pronuclei only of type A, and the homozygous a/a cell produces only a pronuclei. The zygotic nucleus in each conjugant must, therefore, be heterozygous, A/a, and all the nuclei derived mitotically from the zygote should also be heterozygous.

If a population of heterozygous cells should be crossed with another population of heterozygotes (of another mating type) a classical $1:2:1$ ratio is obtained (Fig. 7–5). Although different genotypes are produced in the cross, differences do *not* appear among the products of a single conjugating pair, because the fertilization nuclei in the two members of a pair are identical, and the subsequent nuclear divisions are mitotic.

←——————————————————————————————

conjugant moves into the paroral cone; the macronuclei form skeins. (*4*) The haploid pronuclei divide mitotically to provide a migratory and a stationary nucleus in each conjugant; the macronuclei are now fragmented. (*5–7*) The migratory nuclei are exchanged, approach the stationary nuclei, and fuse with them to form the diploid zygote nuclei (synkarya). (*8*) The synkaryon divides twice mitotically, placing two diploid nuclei at each end of the exconjugants. (*9*) The nuclei at the posterior end begin to develop into macronuclei and move to the center of the cell. (*10*) At the first postconjugal cell division both micronuclei in each exconjugant divide, but the new macronuclear primordia are separated into the daughter cells. (*11*) These first fission products, the origins of the karyonides, now have the normal vegetative nuclear equipment. (*12*) At subsequent cell divisions both the micronuclei and the macronuclei divide; the fragments of the old macronucleus are assorted and eventually disintegrate. Reprinted, by permission, from Beale, G. H. 1954. *Genetics of Paramecium aurelia*, Cambridge University Press. Fig. 1, p. 27.

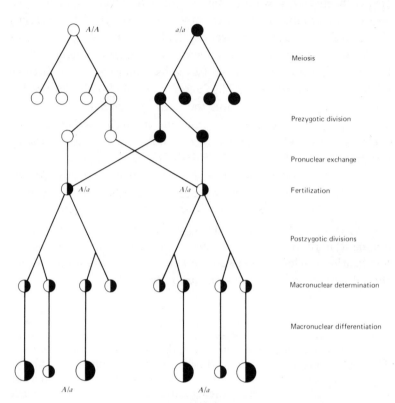

Fig. 7-4. Genetic consequences of crossing homozygous strains of *Tetrahymena thermophila*. The micronucleus in one strain undergoes meiosis to yield haploid nuclei of type *A*; the micronucleus in the other strain produces haploid products of type *a*. One randomly chosen haploid nucleus in each cell divides mitotically to produce identical haploid products: *A* in the first strain, *a* in the other. Pronuclear exchange and fertilization establish identical heterozygous synkarya in the mates. The postzygotic nuclear events establish two heterozygous macronuclei and one heterozygous micronucleus in each exconjugant.

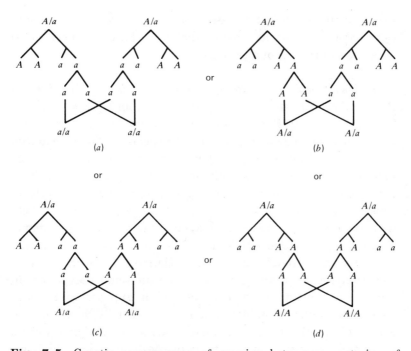

Fig. 7-5. Genetic consequences of crossing heterozygous strains of *Tetrahymena thermophila*. Only the prezygotic events need to be considered. Meiosis yields two A and two a haploid nuclei in each conjugant, with equal probability of giving rise to the pronuclei. The nucleus chosen in both mates may be a (event a) or A (event d), or nuclei of different genotypes may be chosen in the two conjugants (events b and c). The exconjugants are *always* alike in genotype, regardless of which prezygotic events occur. The genetic result is that one quarter of the *pairs* are a/a, one quarter are A/A, and one half are A/a.

B. Conjugal Variations

The sequence of events in conjugation varies from ciliate to ciliate, and even within the same species. Some strains have alternative forms of conjugation, and the events are sometimes subjected to experimental modification. Because these cytogenetic permutations have been used as a major analytical device, we need to discuss a few of the major alternatives.

1. Variations in numbers of nuclei. We note earlier that the number of nuclei of each type varies from species to species. The species of the *P. aurelia* complex characteristically have two micronuclei and one macronucleus; the species of the *P. caudatum* and *P. bursaria* complexes have one micronucleus and one macronucleus; the *P. multimicronucleatum* species, as might be guessed, have several micronuclei, and the number is not perfectly constant, even though all clones regularly begin their life cycles with four. Such variations in micronuclear number are common among the ciliates, and they do affect the course of conjugation, but only in a trivial way. In conjugation, *all* the micronuclei characteristically undergo meiosis, though some exceptions are known. A multimicronucleate cell then appears to have an enormous number of haploid micronuclei. Usually, however, only one of this number of haploid nuclei finds shelter and produces the pronuclei; all the remaining nuclei become relics which are eventually resorbed. (One should not make categorical statements, perhaps, even in a primer. In some strains or species of ciliates the stationary pronucleus is sometimes derived from another haploid nucleus and is not the sister of the migratory pronucleus.)

Numerical variations in the macronuclei may also complicate the conjugal maneuvers. In Euplotes species only one postzygotic division occurs before the macronucleus is induced; the exconjugant clone is composed of a single karyonide. In species of the *P. multimicronucleatum* (or *P. caudatum*) complex, in contrast, the fertilization nucleus usually divides three times instead of twice before the new macronuclei are induced. Instead of two macronuclear anlagen developing, four are produced in each conjugant. These four anlagen cannot be separated completely at the first cell division after conjugation, and the process requires two divisions. Consequently, each exconjugant yields four karyonides and each synclone consists of eight karyonides. Variations in the numbers of postzygotic nuclear divisions are known even within a species (*P. primaurelia*) and have been used to experimental advantage. In all these cases, however, the standard vegetative condition with one macronucleus is quickly reached and subsequently maintained.

Some ciliates, particularly among the hypotrichs, regularly maintain two macronuclei in vegetative cells, and some macronuclei become lobulated to appear like a string of sausages or a branched tree. But these more or less discrete macronuclear elements coalesce prior to cell division and are subsequently reconstituted after division. The form changes and numerical variation probably reflect physiological requirements.

2. Variations of cytoplasmic union. The "standard" conjugation pattern described above provides only membrane fusion between the mating cells and a strict equivalence of function of the mating cells. Yet some ciliates, particularly those with sessile forms, have developed structural and functional distinctions between the two partners (Fig. 7–6). In such cases one may refer to *anisogamonts* in contrast to the more usual *isogamonts*. The microconjugant is taken into the macroconjugant and ceases to be a separate entity. Little genetic work is available on the peritrichs and other groups that produce microconjugants and macroconjugants, and we do not dwell on the phenomenon here.

Even when the conjugating cells are functionally similar and of the same size, variations as to the intimacy of union can be observed. In many hypotrichs, for example, the mating cells fuse completely and have continuous endoplasm. In some cases the conjugating pair is never reconstituted into separate exconjugants, and the process is called *hologamy*. More commonly, the fused mates are eventually separated, but after an interval of profound cytoplasmic union.

From an experimental point of view, the degrees of cytoplasmic intimacy are interesting because they can be manipulated. In *P.*

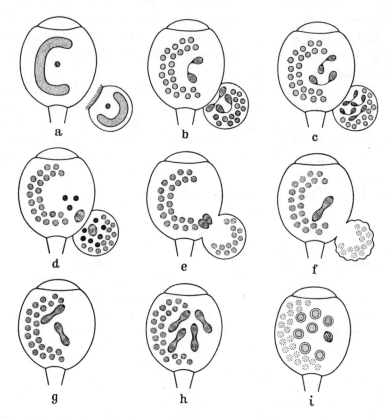

Fig. 7-6. Conjugation in a peritrich ciliate. Here the conjugants are structurally and functionally distinct. One is large and sessile (the macroconjugant) and one is small and motile (the microconjugant). These "anisogamonts" (a) become agglutinated, (b,c) undergo meiosis, and (d,e) undergo fertilization, but the process is not reciprocal. The synkaryon gives rise to a new system of nuclei (f-i), while the old macronucleus fragments and disintegrates. Reprinted, by permission, from Grell, K. G. 1973. *Protozoology,* Springer-Verlag. Fig. 201, p. 208.

tetraurelia, the conjugating cells regularly separate from a super-
ficial contact 6 hours after union (at 27 °C). A few pairs, however,
may fail to separate and are seen to have a cytoplasmic bridge of
variable width connecting the mates. The more extensive bridges
may persist for several hours, during which cytoplasmic particles
may be seen passing through the bridges. In rare cases the bridges
include the entire right ventral surfaces of the cells and they never
achieve separation, but reorganize slightly as *double animals* and
breed true in this form through subsequent cell divisions (Chapter
10). Although bridges and double animals occur spontaneously,
they may be selected regularly after treating conjugating cells with
immobilizing antiserum prior to separation. One can control, in
effect, the degree of *cytoplasmic mixture* and ascertain the effects
of such mixtures on the characteristics of the clone.

3. Variations in nuclear behavior. Numerous variations in
nuclear behavior have been reported, and several of these are
important from a genetic and experimental perspective. The
following list is not exhaustive, but it illustrates the range.

a. *Cytogamy.* In *P. tetraurelia,* as well as other species of the
complex, paired cells are ordinarily attached in two regions; they
may be held together firmly in the anteriorventral area (the *hold-
fast* region), but fail to achieve a firm union in the midventral area
(*paroral* region) where pronuclear transfer occurs. The migratory
pronuclei fail to be transferred, for this or other reasons, and
instead are redirected to a fertilization with the stationary nucleus
in the same cell. Since this stationary nucleus is a sister nucleus
only one mitotic division removed, the fertilization nucleus formed
is necessarily homozygous. Although both mates become
homozygous, they do not necessarily become homozygous for the
same genetic determinants. Undetected cytogamy is a hazard in
genetic analysis and can lead to an unwarranted interpretation of
"cytoplasmic inheritance," if systematic differences between
"exconjugant clones" are found. Cytogamy can in some instances

be brought under experimental control and can be a useful means of inducing immediate homozygosity.

Induced cytogamy has recently been achieved with *T. thermophila* by transferring synchronized populations of mating pairs into a medium with increased ionic strength. To be effective the change of medium must occur at a critical time, near the time of the third prezygotic nuclear division. Over 30% of treated pairs may undergo cytogamy instead of conjugation under these conditions.

b. *Autogamy.* Autogamy, as we note in Chapter 7, is a regular occurrence in the species of the *P. aurelia* complex and in one species of Tetrahymena. The basic nuclear behavior is that observed in cytogamy, except that it occurs in single cells without association with cells of another type (Fig. 7-7). The genetic consequence of autogamy is, of course, homozygosity.

c. *Macronuclear regeneration.* The old macronucleus is destroyed in the normal process of conjugation. In Tetrahymena it rounds up, loses its granular appearance, becomes smaller in size, and is completely destroyed before the first cell division. In Paramecium the loss is not so abrupt. The macronucleus develops *skeins* and *lobulations* and breaks down into 30–40 fragments. The disintegration does not proceed directly, however, for the fragments are still active in RNA synthesis and they persist for at least five or six cell divisions after conjugation if the exconjugants are well-fed. Because they do not divide in the presence of a new macronucleus, the number of fragments per cell is reduced by half at each cell division until cells are produced with no fragments. The much more compound Paramecium may require a more gradual transition between the generations than does Tetrahymena and may achieve it by this dilution of a functional old genetic system while the new genetic apparatus is being elaborated.

Under certain circumstances, the fragments of the Paramecium macronuclei may be "rescued" from their normal fate by removing them from the inhibitory influence of the new macronucleus

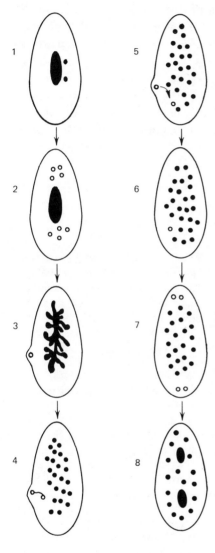

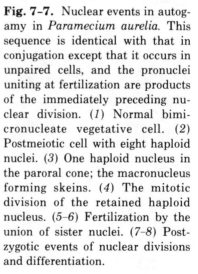

Fig. 7-7. Nuclear events in autogamy in *Paramecium aurelia.* This sequence is identical with that in conjugation except that it occurs in unpaired cells, and the pronuclei uniting at fertilization are products of the immediately preceding nuclear division. (*1*) Normal bimicronucleate vegetative cell. (*2*) Postmeiotic cell with eight haploid nuclei. (*3*) One haploid nucleus in the paroral cone; the macronucleus forming skeins. (*4*) The mitotic division of the retained haploid nucleus. (*5–6*) Fertilization by the union of sister nuclei. (*7–8*) Postzygotic events of nuclear divisions and differentiation.

(Fig. 7-8). When no macronucleus develops, or when a defective macronucleus forms the fragments of old macronucleus begin to grow. At cell division the number of fragments is reduced, but the size of the fragments increases, until each cell has only one fragment, but that fragment is the size of a normal macronucleus and is the normal functional macronucleus of the cell.

This process of macronuclear regeneration is commonly encountered in Paramecium when the fertilization nucleus is defective for genetic reasons. Macronuclear regeneration (MR) is therefore, like cytogamy, a hazard to the unwary geneticist. It can

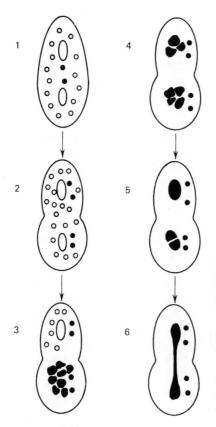

Fig. 7-8. Macronuclear regeneration in *Paramecium aurelia*. (*1*) Normal exconjugant or exautogamous cell with two new macronuclei, two micronuclei, and many (30–40) fragments of the old macronucleus. (*2*) Normal first postzygotic cell division, with micronuclei dividing, macronuclear fragments being assorted, and the new macronuclei being distributed to the daughter cells. (*3*) Abnormal second postzygotic cell division (or later) at which new macronucleus fails to divide; one daughter receives only fragments. (*4–5*) Subsequent cell divisions in lineage with only fragments. Fragments become larger and few, until the single remaining fragment is a normal sized macronucleus that divides when the cell divides. (*6*)

also, however, be a powerful tool, particularly because it can be induced at will. Temperature shocks, for example, delivered to late conjugants may abort the new macronuclei completely and allow the macronuclear fragments to grow instead. Sometimes the temperature shocks only delay the growth of the new macronuclei for a few cell divisions, causing some subclones to be nucleated by fragments and others by new macronuclei. Macronuclear regeneration may also be induced in high frequency without temperature shocks by the action of any one of a number of recessive genes, e.g. *am*. By using appropriate genetic markers, one may identify the branches of a *vegetative pedigree* that contain regenerated fragments and hence assess the effects of nuclei of different origin assorting in a common cytoplasm.

Macronuclear regeneration is also a means of constructing *heterokaryons*. Because the new macronucleus is more sensitive to environmental insults than is the micronucleus, the regenerated macronucleus may coexist in a cell with a genetically different micronucleus. We refer earlier to a conclusion concerning nuclear functions derived from such heterokaryons.

Macronuclear regeneration is not limited to the genus Paramecium, but also occurs in Tetrahymena. It can also be induced by environmental or genetic manipulation and has similar consequences. In Tetrahymena, however, the macronucleus never fragments, and the entire old macronucleus is retained. For this reason the comparable process in Tetrahymena is called *macronuclear retention*.

d. *Genomic exclusion.* Because the micronucleus has a limited vegetative role in ciliates, grossly defective micronuclei may be carried along with relatively little effect. At conjugation, however, the micronuclear defects become manifest and may lead to alternative kinds of reorganization. One particular use of lines with defective micronuclei is the production of homozygotes. In *Tetrahymena thermophila* a cross between a normal cell and a grossly defective cell (called a* or "star" cell) results in pronuclei being formed only in the normal cell. Nuclear transfer usually occurs in the normal way, but unidirectionally—and perhaps after a delay. Each exconjugant thus receives a haploid micronucleus rather than a fertilization nucleus. In most such cases, the haploid nucleus diploidizes, but fails to make the normal developmental schedule. Macronuclear retention occurs and a heterokaryon is formed with a homozygous micronucleus (Fig. 7–9). This heterokaryon may be converted into a homokaryon again by crossing it and replacing the old macronucleus. In this way the entire genome of the original "star" cell is replaced by that from the normal cell. First round genomic exclusion, by creating a

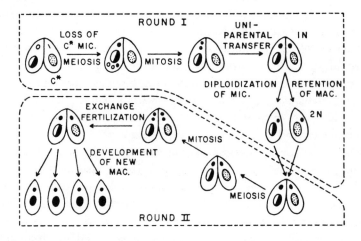

Fig. 7–9. Genomic exclusion in *Tetrahymena thermophila*. This process occurs regularly in crosses between normal diploid strains and certain strains with defective micronuclei. It usually requires two pairing events, "round 1" and "round 2." In round 1 the defective strain (C*) loses its micronucleus, but receives a pronucleus from the normal mate. The micronuclei in both mates diploidize, in an incompletely understood way, but they do not give rise to a new macronucleus. At a subsequent conjugation, however, the now homozygous micronuclei in both cells behave normally and give rise to functional macronuclei. Reprinted, by permission, from Bruns, P. J., T. B. Brussard and A. B. Kavka 1974. Positive selection for mating with functional heterokaryons in *T. pyriformis*. *Genetics*, **78**, 831–841, Fig. 1, p. 835.

heterokaryon, makes it possible to examine the behavior of identical micronuclei in different cytoplasmic environments, under the influence of different macronuclei. The diploidization associated with the process also yields instant homozygosity and is a valuable method for obtaining recessive mutants or for rendering wild strains homozygous.

The sequence of two episodes of mating before the micronuclear genes of one strain become expressed as homozygotes in the other is the most common one, but sometimes the haploid pronucleus both diploidizes and gives rise to new macronuclei during a single conjugation cycle. This short circuit genomic exclusion occurs in only a few percent of the conjugating pairs, but mass selection techniques have been developed with drug sensitive macronuclei and drug resistant micronuclei; the only survivors of crosses in the presence of the drugs are the conjugants undergoing short-cut genomic exclusion to express their drug resistant markers. These heterokaryons are useful to guard against macronuclear retention wherever mass matings are used in genetic analysis.

One additional cytogenetic maneuver was devised recently to produce first polyploid variants and then secondarily aneuploids of various composition. The technique derives from observations on triple conjugations, unusual associations of three cells united in conjugation. Two basic kinds of triplets occur in *T. thermophila.* The most common kind found when three or more different mating types are mixed is a symmetrical association of three different mating types (Fig. 7–10). Each cell undergoes meiosis and develops pronuclei; nuclear transfer occurs *to* the mate on the right and *from* the mate on the left. Diploid zygote nuclei are thus established in all three cells and these proceed to develop normal macronuclei.

The rarer, but more useful, triplets are those occurring primarily in mixtures of only two mating types. Two cells of a single mating type are associated with another of a different mating type, but are unable to interact with each other. The attachments,

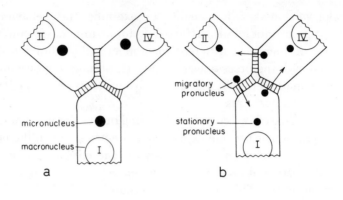

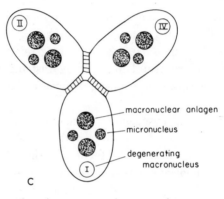

Fig. 7-10. Summary of major cytogenetic events in symmetric triplets composed of three different mating types. (*a*) Premeiotic stage when each cell is fused to both other cells, the left side of the fusion membrane associated with one cell and the right side with the other. (*b*) Stage of pronuclear transfer at which the migratory nucleus of each cell moves into the cell on its right to fertilize the stationary nucleus. (*c*) Stage prior to separation when each conjugant contains two new micronuclei and two new macronuclei—of the same sizes in all cells, but of different genotypes in certain combinations of parental strains. The mating types of the parents are indicated by the roman numerals on the macronuclei, which are being resorbed in *c*. Reprinted, by permission, from Preparata, R. M. and D. L. Nanney, 1977. Cytogenetics of triplet conjugation in Tetrahymena: origin of haploid and triploid clones. *Chromosoma,* **60,** 49–57. Fig. 3, p. 55.

116

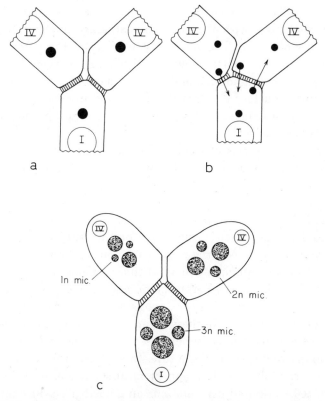

Fig. 7-11. Summary of major cytogenetic events in asymmetric triplets, commonly composed of only two mating types, I and IV. (*a*) Premeiotic state, all cells touch, but firm unions are formed only between cells of different type. (*b*) Stage of pronuclear transfer, during which both cells of type IV transfer a pronucleus to the same cell of type I (the "bridge" cell), which is able to reciprocate with only one of the type IV cells, leaving the other type IV (the "third mate") without an external contribution. (*c*) Late stage following fertilization and two postzygotic divisions. The haploid third mate has small micronuclei and macronuclei; the bridge cell has large, presumably triploid, nuclei of both kinds. Reprinted, by permission, from Preparata, R. M. and D. L. Nanney, 1977. Cytogenetics of triplet conjugation in Tetrahymena: origin of haploid and triploid clones. *Chromosoma,* **60,** 49–57. Fig. 3, p. 55.

moreover, are often somewhat asymmetric (Fig. 7–11). Both of the attached cells often contribute a pronucleus to the common mate, where they fuse with the resident stationary nucleus to give a triploid synkaryon. The migratory nucleus of the common cell goes to one of the mates, yielding a diploid synkaryon, but one of the mates receives nothing and remains haploid. For some reason this haploid nucleus does not diploidize, but divides in synchrony with the zygote nuclei of the other conjugants and produces functional macronuclei.

The euploid series of clones, plus the "star" clones with nonfunctional micronuclei, may be used to generate special strains with various chromosome combinations useful for experimental manipulations.

All the variations in nuclear behavior thus far reported in the ciliates have not been mentioned, for this subject alone could occupy an entire book the size of this one. Instead attention is directed to the principal events in those organisms most used in genetic investigations—Paramecium and Tetrahymena—so that the breeding studies using the organisms may be comprehensible. Some of the variations in these organisms are elaborated, because the variations have already been used in analysis and will be used even more frequently in the future. In any case, the variations are not so formidable as they appear at first glance, once they are recognized as limited permutations on a basic procedure invented long ago and resistant to fundamental change.

SUMMARY

The basic cytogenetic event in the life cycle is conjugation. Conjugation is a process of reciprocal fertilization between (usually) cells of different mating type. The pronuclei exchanged are produced following meiosis and a postmeiotic mitotic division; this mitotic division assures the identity of the migratory and stationary nuclei and consequently the genetic identity of the exconjugant cells. The fundamental similarities in the cytogenetic patterns argue strongly for the common origin of the ciliates.

The basic cytogenetic patterns are modulated, however, in response to ecogenetic demand or experimental exigency. Some inbreeding species undergo autogamy, a relatively minor pattern variation in which the pronuclei of a single cell unite to establish a homozygous nuclear apparatus. Other variations are also described, and some of these form the basis of a powerful genetic technology.

RECOMMENDED READING

Allen, S. L. 1967. Genomic exclusion: a rapid means of inducing homozygous diploid lines in *Tetrahymena pyriformis,* syngen 1, *Science (Wash. DC),* **155,** 575–577.

Grell, K. G. 1967. Sexual reproduction in protozoa. *Res. Protozool.,* **2,** 147–213.

Maupas, E. (1889). La rajeunissement karyogamique chez les ciliés. *Arch. Zool. Exp. Gén.,* **7,** 149–517.

Orias, E. and E. P. Hamilton. 1979. Cytogamy, an inducible alternate pathway of conjugation in *Tetrahymena thermophila. Genetics,* **91,** 657–671.

Raikov, I. B. 1972. Nuclear phenomena during conjugation and autogamy in ciliates. *Res. Protozool.,* **4,** 147–289.

Sonneborn, T. M. 1947. Recent advances in the genetics of Paramecium and Euplotes. *Adv. Genet.,* **1,** 263–358.

Wenrich, D. H. 1954. Sex in protozoa. In *Sex in Microorganisms* (D. H. Wenrich, Ed.), American Association for the Advancement of Science pp. 134–265.

8

The Ciliate
Macronucleus

THE CILIATE MACRONUCLEUS IS AS YET AN ENIGMA, but it is beginning to yield to analysis. Or perhaps we should say that *they* are beginning to be understood, because evidence is accumulating that the superficially similar macronuclei of different ciliate groups are organized according to different principles. Macronuclear phenomena observed in one ciliate species may never be encountered in another. Here we discuss a few studies on ciliate macronuclei that illustrate their variety and their experimental utility.

A. Chromatin Diminution

Chromatin diminution has long been reported as an unusual accompaniment of development. In a wide assortment of organisms including some crustacea, insects, and nematodes, a substantial fraction of the germinal DNA content is eliminated from the cells destined to yield the somatic tissues; all the DNA sequences of the species are retained only in the cells that give rise to the germ cells in the following generation.

In a similar way, the development of the macronucleus in hypotrichs (Euplotes, Stylonychia, etc.) involves an elimination of many of the DNA sequences composing the micronuclear chromosomes. The "diminution" may be associated with a spectacular developmental display. Polytene chromosomes, at least superficially similar to the polytene chromosomes in the Drosophila salivary gland, are elaborated in the macronuclear anlagen. These polytene chromosomes are short-lived, however, and quickly disintegrate. Much of the DNA content of the macronucleus is lost, and then a new cycle of DNA replication begins. The new DNA, however, includes a fraction of the original micronuclear DNA sequences. Molecular hybridization of macronuclear and micronuclear DNA sequences shows that as little as 1% of the micronuclear DNA sequences may be represented in the mature macronucleus in some species.

123

The phenomenon of chromatin diminution in ciliates (as in other organisms) provokes questions concerning the role of the excess DNA in the germ line. No satisfactory answers are yet available. We should note, however, that species of ciliates with chromatin diminution generally have larger amounts of DNA in the micronucleus than do species without chromatin diminution. The number of unique sequences in the different kinds of organisms may be very similar and of the same order as is found in Drosophila. The excessive germinal DNA may be repetitive and, at least in an evolutionary time scale, dispensable.

We should emphasize that the dramatic macronuclear changes and the sharp differences between macronuclear and micronuclear composition are observed only in some ciliates; in the most extensively studied species such as the *P. aurelia* and *T. pyriformis* complexes the loss of DNA sequences is limited to 10% or less. Nevertheless, more chromatin diminution may occur in *P. bursaria,* and preliminary data suggest that much rearrangement of the chromatin may occur during the transformation of the micronuclei into macronuclei, even in the apparently more conservative organisms.

B. Quantitative DNA Regulation

Chromatin diminution is a highly selective developmental process responsible for qualitative genetic differences between germ line and somatic nuclei. Another process, called *chromatin extrusion,* occurs in some ciliates and may have a quantitative regulatory significance. We return to chromatin extrusion after trying to place it in perspective. At an ordinary cell division in Tetrahymena, the micronucleus divides, the macronucleus divides, and the cell divides, but micronuclear and macronuclear DNA syntheses do not necessarily occur at the same time. Micronuclear DNA synthesis, for example, occurs immediately after micronuclear division in Tetrahymena, coincidentally with cytokinesis.

Macronuclear DNA synthesis, in contrast, usually begins about midway through the cell cycle and continues to nearly the end.

The micronuclear and macronuclear replication cycles differ not only in their timing, but also in their mechanisms and accuracy of division. That is, micronuclear divisions are quantitatively precise and produce exactly equivalent daughter nuclei, but macronuclear divisions are "amitotic" and produce only approximately equal daughter nuclei. The average difference between the DNA amounts of daughter nuclei is 8%, and differences as high as 2 to 1 are not uncommon. Gene mutations (in Paramecium) are known that greatly increase the eccentricity. Macronuclear DNA replication in Tetrahymena, but not in Paramecium, is an all-or-none process, so that inequalities produced at one cell division are perpetuated into the next cell cycle and may be augmented by a subsequent unequal division.

Yet, the macronuclear DNA content in a culture of Tetrahymena does not vary so widely as would be expected on the basis of the unequal divisions. The regulation of macronuclear DNA content apparently involves two processes. The first of these is the insertion of a second macronuclear DNA replication cycle within a cell cycle. When the DNA content falls below some critical level, cytokinesis is deferred, and a second replication cycle is undertaken. This procedure prevents the accumulation of cells with very low macronuclear content, but the way the cell senses its own DNA content has not been discovered.

Cells that receive an unusually large amount of macronuclear material also require a method of compensation. Since macronuclear and cell division may be separated, a process opposite to that employed for cells with small macronuclei would be a plausible means of regulation. Indeed, under some circumstances cells do undergo two cell divisions (and macronuclear divisions) with only a single macronuclear S phase. Such atypical divisions may be induced when cytokinesis has been inhibited, as by temperature shocks, without interfering with DNA synthesis. The excessive DNA is restored to normal levels by a series of divisions without macronuclear S phases. Some indication is available, moreover, that such uncoupled cell cycles occur regularly at some point in the cultural growth cycle for some strains.

It is within this context, perhaps, that chromatin extrusion should be viewed. At the end of macronuclear division the macronucleus takes the form of an hourglass with a progressively more narrow waist. Within this membrane surrounded channel one may often observe a bulge of chromatin, which is eventually separated from both daughter nuclei and comes to lie near the fission furrow of the cell at cytokinesis. The *extrusion bodies* are of variable size, and most are resorbed part way through the cell cycle. Although particular kinds of DNA sequences might be sequestered in the extrusion bodies, a more likely explanation is that their production compensates only quantitatively for somewhat high DNA levels in some cells and that their content is not specific. Computer simulation suggests that chromatin extrusion may not even compensate quantitatively.

The total DNA content of the ciliate macronucleus is in any case restrained within fairly narrow limits, not by a differential regulation of DNA synthesis (at least in Tetrahymena), but by compensatory devices that reduce the DNA amount when it reaches certain levels and increases the amount when it falls too low. How the macronucleus maintains its genetic integrity in the face of these mechanisms is a problem we return to shortly.

C. Genetic Amplification

Another kind of nuclear alteration has been reported, in both Tetrahymena and Paramecium. Some DNA sequences present in small numbers in the micronucleus are represented proportionately much more in the macronucleus.

The best established instance of DNA amplification in ciliates (indeed the only known instance) involves the sequences that encode the ribosomal RNAs. The macronucleus in Tetrahymena contains about 45 times as much total DNA as the micronucleus. The micronucleus contains only one copy of the genes for the ribosomal RNAs per haploid genome (two per diploid micronucleus), but the macronucleus contains over 12,000 copies of the ribosomal cistrons, or about 300 copies per haploid equivalent of DNA.

The additional rDNA cistrons are, moreover, organized in a different way in the macronucleus and the micronucleus. In the micronucleus the rDNA is associated in a normal way with other DNA sequences in a chromosome and consists of a single conventional linear sequence. In the macronucleus the rDNA is not integrated at all with other sequences and occurs as isolated palindromes—adjacent sequences of reversed polarity (Fig. 8–1). No integrated copy of rDNA can be discovered in the entire macronucleus, even though sensitive tests could reveal it. The rDNA may also replicate out of phase with the bulk of the macronuclear DNA.

The amplification of the rDNA has also been observed in Paramecium. The paramecia are larger and more compound than the tetrahymenas. *P. tetraurelia* has enough DNA for about 860 micronuclear haploid equivalents, but the rDNA of the macronucleus is disproportionately augmented, and as in Tetrahymena, the rDNA molecules are disjoined from the other DNA sequences, though they do not take the form of palindromes, but rather of tandem repeats. DNA amplification of the rDNA may be a general feature of the ciliate macronucleus, even as it is a common

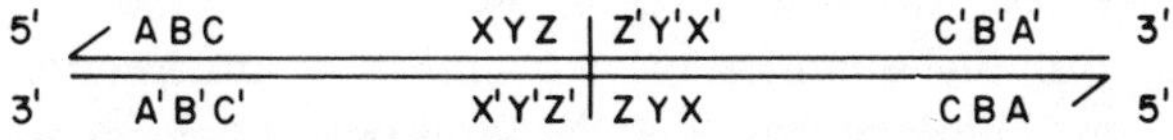

Fig. 8-1. Structure of a linear rDNA molecule from Tetrahymena. The molecule has an axis of rotational symmetry: the two polynucleotide strands are identical and each is self-complementary about its center. Reprinted, by permission, from Gall, J. G., K. Karrer, and M-C. Yao, 1977. The ribosomal DNA of Tetrahymena. In *The Molecular Biology of the Mammalian Genetic Apparatus* (P. Ts'o, Ed.), *Elsevier*, pp. 79–95. Fig. 1, p. 80.

process in the early development of multicellular organisms. The rDNA amplification of ciliates may be distinctive, however, in the form of the disjoined cistrons.

The amplification of rDNA is a solution to an important biological problem: the achievement of a large synthetic capacity while maintaining strict evolutionary control of the synthetic machinery. It is not a commonly adopted solution, however, and is not even employed for all components of the synthetic system. Transfer RNAs and 5 S RNAs for example, are not amplified in Tetrahymena, and no convincing evidence is yet available for amplification of other genomic components in ciliates. One possible explanation for the rarity of DNA amplification (at least disjunctive amplification as opposed to integrative amplification) is that a special system is required to maintain the appropriate numbers of the special genomic elements. This system must be capable of measuring the number of the amplified components and of linking that measurement to their replication. Perhaps such systems are too expensive to be established for any but the most essential cellular components.

D. Phenotypic Assortment

Although the studies briefly summarized above begin to provide some understanding of the ciliate macronucleus, they do not explain its organization. The students of paramecia have generally been impressed with the phenotypic stability of clones, which contrasts with the quantitative variability of macronuclear division. Clones heterozygous for co-dominant alleles continue to express both alleles. Clones heterozygous for dominant alleles continue to express the dominant alleles. Even when the macronucleus is fragmented into 30–40 pieces and regenerated from individual pieces, each fragment yields a lineage expressing the original genotype.

To explain this genetic conservatism, the idea of diploid subnuclei was proposed. The macronucleus was thought to consist of diploid sets of chromosomes associated by unexplained connections, but not enclosed within separate nuclear membranes. The division of the macronucleus was thought of as a distribution of the "subnuclei" in only approximately equal numbers to the two daughter cells. Even when macronuclear fragmentation was induced, the 860 haploid equivalents in DNA could provide an average of over 10 diploid sets to each of 40 fragments. Regardless of the differences in the numbers of subnuclei, in daughters or in fragments, every cell would be provided with proper *proportions* of the genetic elements. Genetic balance would be preserved, and the continuity of genotype would be assured.

The interpretation is still defensible for Paramecium, for no cases of genetic assortment have been established. The search has been extensive, but the level of compoundness is so high and the number of divisions is so small that the negative evidence is not compelling.

For Tetrahymena, however, a new phenomenon has been discovered. Particularly in *T. thermophila,* nearly all heterozygotes manifest *phenotypic assortment.* Clones heterozygous for codominant alleles produce sublines stably manifesting only one of the alleles. Clones heterozygous for a dominant gene initially express the dominant alleles, but eventually produce sublines manifesting the recessive allele. The evidence of phenotypic assortment implies an underlying genetic assortment. Because genetic assortment is difficult to imagine with diploid subnuclei, alternative interpretations of the Tetrahymena macronucleus have been considered.

To understand the genetic behavior of the Tetrahymena macronucleus and its interpretation, we must consider briefly the general characteristics of an assorting system of this type. Let us consider that a container (the nuclear envelope) surrounds N items (genetic elements) capable of being sorted into two classes (say black and white). We imagine that each item "replicates," that is each black object yields two black objects and each white object yields two white objects, providing a total of $2N$ items (Fig. 8-2). Now consider that the population is thoroughly mixed and divided into precisely equal portions (daughters). In each daughter the process is repeated: replication, mixture, and partition. What are the consequences of this activity?

Inequalities of daughter composition can occur at any division and these inequalities are increased by sampling variation at subsequent divisions until some daughters contain mainly white objects and others mainly black objects. Eventually daughters will be produced with only black or white objects and these will "breed true" thereafter, unless black objects can be converted into white objects, or vice versa. As the process is continued, the fraction of

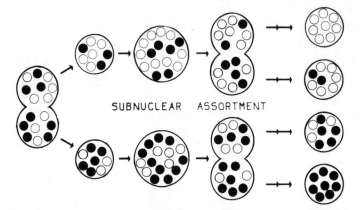

SUBNUCLEAR ASSORTMENT

Fig. 8-2. Model of assortment of genetic elements in the macronucleus of Tetrahymena. Two classes of determinant are present; these replicate true to type, but are distributed at division randomly with respect to kind. Pure nuclei (pure black or pure white) are produced at a rate determined by the total number of elements present. Reprinted with permission from Nanney, D. L. 1963. Aspects of mutual exclusion in Tetrahymena. In *Biological Organization at Cellular and Supercellular Levels*. (R. J. C. Harris, Ed.) Academic Press, pp. 91–109. Fig. 2, p. 95. Copyright by Academic Press Inc. (London) Ltd.

daughters of "mixed" composition becomes progressively smaller until it becomes negligible.

This model of assortment incorporates several assumptions, some of which are known to be incorrect. What are the effects of modifying these assumptions? Before we answer this question we must give brief attention to certain quantitative considerations. The probability that a daughter will be produced with only one kind of item is related to the number of items (N) in the container. If the container holds 1000 items, very few daughters will have pure compositions, but if the container has only 10 items, many more daughters from mixed sources will become pure. The precise relationship between N and R_f (the rate of fixation) is described by the Schensted formula, $R_f = 1/(2N - 1)$, and was arrived at first by computer simulation. The rate of fixation has been experimentally determined for a number of different examples of phenotypic assortment and has been found to be 0.0113 per cell division. When R_f is 0.0113, $N = 45$. Thus within its assumptions the model proposes a macronucleus with 45 "items" after cell division or 90 items before division.

What are the effects of violating the assumptions of the model? We pursue this question only qualitatively. First we note that the model postulates precisely equal division of the elements among the daughter nuclei; yet we report above that the *average* discrepancy is about 8%. Unequal distribution of elements to the daughter cells increases the probability that a pure daughter will be produced. In the extreme instance, for example, one daughter would receive $2N - 1$ elements and the other only 1; the daughter with only 1 element would of course have only one *kind* of element. Distribution between this inequality of numbers and the equality postulated would give a graded probability of fixation. If, instead of estimating fixation rates from the number of elements, we estimate the elements from the rate of fixation, an R_f of 0.0113 might suggest somewhat more that 45 elements, the increase of estimated elements depending on the degree of inequality of distribution of the elements. Computer simulation indicates that slight inequalities at division cause little change in R_f so long as the mean number of elements is constant.

The model suggested that the items are replicated and then thoroughly mixed, so that their distribution to daughters is unrelated to their origin, that is, "sister elements" are randomly distributed. But two kinds of nonrandom assortment can be imagined. Perhaps the elements are not thoroughly mixed after replication, but tend to remain in their original associations. In this case the probability of producing a pure daughter is increased to the degree that sister elements remain associated. Thus if sister elements remain firmly attached for one cell cycle and then became randomized, the calculated number of "assorting units" would be precisely half the actual number of units in the container. To obtain an R_f of 0.0113, N would have to have a value of 90, not 45.

Alternatively, we might imagine that sister elements when replicated tend to move to opposite sides of the container and so come to rest in different daughters more frequently than would be expected by chance. In that case an R_f of 0.0113 could be consistent with a number of units much lower than 45. In the most decisive experiments on this subject, in which Orias and Flacks analyzed the distribution of the sister elements when only one element of a kind was present before replication, no evidence of nonrandomness was found. Therefore, the number 45 seems a reasonable estimate of the number of assorting units.

One special feature of the assortment model requires additional consideration. We began with the assumption of equal numbers of black and white units. Some heterozygotes do behave in fact as if they contained equal numbers of units, but some show considerable eccentricity of output, that is, the pure lines of one type may greatly exceed the pure lines of the alternative type when assortment is complete. What is the effect of inequality of composition on the assortment pattern? Computer simulation shows that the R_f value (the probability of producing a pure subline at a particular cell division from a mixed container) is initially very sensitive to inequality of composition, but that by 20–30 cell divisions it approaches the same value regardless of initial inequalities. The measurement of the definitive assortment rate requires, therefore, waiting through a period of equilibration to compensate for input variations.

The inequalities of input also affect the time at which the first pure assortee is expected in an experiment. Unequal compositions yield pure sublines quickly while equal inputs delay the appearances of pure lines. Indeed, the heterozygotes showing an early appearance of pure types are those with unequal output, but whether all the variations in the time at which assortment begins can be interpreted on this basis is unclear. We return to this question later.

The analysis of phenotypic assortment leads to a model that accounts for most of the phenomena observed. In particular, it indicates a macronucleus composed of about 45 randomly assorting "elements," but it does not define those elements. Certain constraints on the nature of the elements are apparent, however,

1. The elements are probably not diploid subunits. The DNA content of the macronucleus is equivalent to 45 *haploid* sets; the macronucleus does not contain enough DNA for 45 *randomly assorting* diploid units.

2. The elements may be haploid subunits. The random assortment of disjoined chromosomes would be expected to cause genetic imbalance. Some sublines would be expected to have a greater proportion of one chromosomal type and others an unusual frequency of another. One can calculate the probability of death or disability occasioned by such assortment. Again the probability of assortment depends on the compoundness of the system, but at least the less highly compound organisms such as Tetrahymena show none of the signs of genetic imbalance that would be required on the basis of random chromosomal assortment.

The alternative to a haploid set of joined chromosomes is a

separate regulatory system that recognizes and compensates for inequalities of chromosome numbers. Yet the reported all-or-none DNA replication in the Tetrahymena macronucleus leaves little room for such a system.

The notion of a haploid chromosome set meets an obstacle, however, in the behavior of linked genes. If no somatic recombination occurs, linked genes should show their original coupling relations after assortment. But available observations indicate no persistence of linkage patterns when linked genes are somatically assorted. The hypothesis of a haploid chromosome set must be supplemented with a hypothesis of somatic recombination at a sufficiently high rate to scramble genetic markers at moderate distances on the same chromosome.

3. The elements may be subchromosomal genophores. The lack of detected somatic linkage is consistent with macronuclear elements smaller than whole chromosomes, and of course smaller than haploid sets of chromosomes. Furthermore, isolated DNA from Tetrahymena and Paramecium is sometimes obtained in lengths substantially shorter than whole chromosomes. The question remains whether technological improvements will allow the isolation of larger segments that approach the size of chromosomes. The DNA pieces now measured, however, are certainly large enough to accommodate many genes, and the study of somatic linkage of closely linked markers should be interesting.

We should mention that the isolated DNA segments from hypotrich macronuclei appear to be substantially smaller than those in Paramecium and Tetrahymena. Whether this is to be attributed to the potency of their nucleases or to a fundamental difference in structure remains to be determined.

The last word has not been said on the ciliate macronucleus. Indeed this chapter may be the one to age most rapidly. I hope so, for the macronucleus has been a challenge and a barrier to experimentalists for too long. The genetic tools for its dissection have seemed inadequate. Now, however, powerful chemical and physical techniques are being focused on the chromatinic structures. The initial results indicate that the macronucleus is far more than simply a multiplied micronucleus. The "determination" of a micronuclear genome may involve a radical reediting of the nucleic script, as well as a change in the histones and other chromatinic proteins. The macronucleus is an instrument of adaptation invented long ago, modified for special circumstances in many diverse evolutionary pathways. The macronucleus may eventually inform us of flexibilities of genetic deployment we could hardly imagine.

SUMMARY

The compound somatic nucleus of the ciliates is a highly specialized genetic system, almost certainly adapted to different functional requirements in different ciliate groups. In all groups, however, the development of a macronucleus from a micronucleus involves a greater or lesser reorganization of the chromatin content. In some species 90% or more of the DNA sequences in the micronucleus are eliminated. In all species examined the rDNA molecules are edited and amplified.

The sizes of the DNA segments in the macronucleus seem to vary among the species. The hypotrich macronucleus appears to contain chiefly small DNA pieces, no longer than a few genes. The holotrichs generally have larger DNA segments, but the sizes of the pieces and their organization have not been completely analyzed. Many workers believe that some mechanism for maintaining balanced genomes is required, but that mechanism has not been discovered.

RECOMMENDED READING

Ammermann, D., G. Steinbrück, L. von Berger, and W. Hennig. 1974. The development of the macronucleus in the ciliated protozoan *Stylonychia mytilus. Chromosoma,* **45,** 401–429.

Berger, J. D. and H. J. Schmidt, 1978. Regulation of macronuclear DNA content in *Paramecium tetraurelia. J. Cell Biol.*, **76**, 116–126.

Cleffmann, G. 1975. Amount of DNA produced during extra S phases in Tetrahymena. *J. Cell Biol.*, **66**, 204–209.

Gall, J. G., K. Karrer, and M-C Yao. 1977. The ribosomal DNA of Tetrahymena. In *The Molecular Biology of the Mammalian Genetic Apparatus* (P. Ts'o, Ed.), Elsevier, pp. 79–85.

Gorovsky, M. 1973. Macro- and micronuclei in *Tetrahymena pyriformis*: a model system for studying the structure and function of eukaryotic nuclei. *J. Protozool.*, **20**, 19–25.

Nanney, D. L. 1964. Macronuclear differentiation and subnuclear assortment in ciliates. In *The Role of Chromosomes in Development* (M. Locke, Ed.), Academic, pp. 253–273.

Orias, E. and Flacks, M. 1975. Macronuclear genetics of Tetrahymena I. Random distribution of macronuclear gene copies in *T. pyriformis*, syngen 1. *Genetics*, **79**, 187–206.

Polyansky, G. I. and I. B. Raikov. 1976. Polymerization and oligomerization phenomena in protozoan evolution. *Trans. Am. Microsc. Soc.*, **95**, 314–326.

Preer, J. R., Jr. 1976. Quantitative predictions of random segregation models of the ciliate macronucleus. *Genet. Res. Camb.* **27**, 227–238.

Prescott, D. M., K. G. Murti, and C. C. Bostock. 1973. Genetic apparatus of Stylonychia sp. *Nature*, **242**, 576, 597–600.

Rae, P. M. M. and B. B. Spear. 1978. Macronuclear DNA of the hypotrichous ciliate *Oxytricha fallax. Proc. Natl. Acad. Sci. U.S.*, **75**, 4992–4996.

Raikov, I. B. 1976. Evolution of macronuclear organization. *Annu. Rev. Genet.*, **10**, 413–440.

Schwartz, V. 1978. Structur und Entwicklung des Makronucleus von *Paramecium bursaria. Arch. Protistenkd.*, **120**, 255–277.

Williams, J. B., E. W. Fleck, L. E. Hellier, and E. Uhlenhopp. 1978. Viscoelastic studies on Tetrahymena macronuclear DNA. *Proc. Natl. Acad. Sci. U.S.*, **75**, 5062–5065.

Yao, M-C, E. Blackburn, and J. G. Gall. 1978. Amplification of the rRNA genes in Tetrahymena. *Cold Spring Harbor Symp. Quant. Biol.*, **43**, 1293–1296.

Yao, M-C. and M. A. Gorovsky. 1974. Comparison of the sequences of macro- and micronuclear DNA of *Tetrahymena pyriformis. Chromosoma*, **48**, 1–18.

Symposium on the Ciliate Macronucleus. 1979. *J. Protozool.*, **26**, 1–35.

Clonal Aging
and
Temporal Patterns

9

A. Somatic Senescence

The ciliated protozoa are widely reputed to be "immortal," and this opinion has influenced the development of our theories of aging. Even before the beginning of this century students of the protozoa were divided as to whether the protozoan clone is able to multiply indefinitely, or whether a clone has a fixed life span. Extensive studies in the early decades of this century seemed to support the possibility of indefinite multiplication in some species. Woodruff at Yale observed a particular strain of paramecium year after year in laboratory culture and periodically reported on its status. The strain showed no symptoms of senescence; it continued to divide regularly hundreds and eventually thousands of times.

Concurrent with these studies were the first explorations of the replicative ability of cells explanted from vertebrates. Alexis Carrel reported that fibroblasts from a chicken could be maintained in tissue culture, long after the bird from which they were derived would have expired. The protozoan observations, combined with the studies on cell cultures, supported the idea that aging is a property of the multicellular state and does not derive directly from intrinsic multiplicative restrictions on the component cells. Perhaps senescence derives from the limitations on growth imposed by a fixed size, the accumulation of toxins, and the deterioration of nondividing cells. In any case, early theoreticians of senescence tended to focus on systemic properties of the multicellular state, rather than on intrinsic constraints on cellular division.

Both bases for this interpretation have been eroded. Relatively few explanted vertebrate cells are able to grow indefinitely, and these are atypical in their chromosome numbers and surface properties. They are genetically altered and their behavior does not explain the age correlated changes in normal cells. Most normal explanted cells undergo a limited number of cell divisions, characteristic for the species, the age of the donor, and their growth conditions, before they lose the capacity to divide.

The studies on Paramecium similarly show immortality to be a myth, based on a misunderstanding. Woodruff and Erdmann observed that the cells of the "Methusalah" strain periodically slowed down in growth rate and went through complex nuclear changes. They observed macronuclear breakdown and the origin of new macronuclei from the micronuclei. They called this process *endomixis* and believed that it was significant in "rejuvenating" the cell lines. Subsequent studies, both cytological and genetic, however, indicate the micronuclear divisions associated with reorganization are meiotic and that the micronucleus giving rise to the new macronuclei is in fact a zygote nucleus—produced by the union of two pronuclei from the same cell. The periodic reorganizations are autogamous (Chapter 7) and represent a full reinitiation of a life cycle from an act of fertilization.

The periodic autogamies in these paramecia invalidate Woodruff and Erdmann's intimations of immortality, but do not of course establish that paramecia have finite life cycles. This discovery was achieved by Sonneborn, who exploited the fact that autogamy requires starvation and determined that these paramecia were not able to continue growth indefinitely in the presence of abundant food. After 250–300 cell generations, all sublines of a clone gradually decline in growth rate and eventually cease to divide altogether.

Thus these paramecia (of the *P. aurelia* complex) show a limited life cycle; their capacity to divide is lost after a specific number of cell divisions. They may be said to show *somatic senescence* because they have a fixed vegetative span. If cells are allowed to undergo autogamy while still dividing vigorously, they have a high probability of producing "young" and vigorous new clones. With increasing age the probability of vigorous and viable progeny declines, until cells with advanced symptoms of senescence are usually incapable of producing normal offspring. This loss of capacity to produce viable sexual progeny is called *germinal aging,* and in paramecia it is commonly associated with somatic senescence. Yet the kinds of age correlated effects are not entirely coordinated. Some relatively young lineages have defective progeny, and some of the offspring of lines that have somatically aged (at least in early stages) are able to grow normally. We return to consideration of germinal aging later.

The basis of somatic aging is difficult to determine. As cells begin to decline in division rate, many changes occur in their molecular composition and cellular activities. The rate of formation of food vacuoles drops off, the ability to repair ultraviolet damage declines, and the total amount of DNA per cell diminishes. Perhaps most of the features of the cell that can be monitored would be found to be altered in some way if careful studies were carried out. But the mechanisms underlying these changes are difficult to assign. Causes and consequences, primary and secondary effects are hopelessly entangled.

Yet, certain general features of somatic senescence are well established. The first is that the changes are "programmed," in the sense that they occur in all sublines of a clone within a definite period of time. The regularity of the program does not exclude chance events as the basis for senescence, but the chance events would have to be reasonably frequent and would have to be "organized," as are atomic disintegrations or falling sand grains in time measuring devices.

A second generalization concerning somatic senescence in ciliates is that it is modifiable. The evidence for this conclusion is somewhat indirect and is based on comparative consideration. The length of the life span is highly variable among the ciliates. Some species characteristically undergo 250–300 cell divisions before becoming senile. Others may have a life span of 450–500 cell divisions. Still others may be able to grow much longer, and some species are probably somatically immortal. These variations in life span are not, moreover, random and meaningless variations, but are an integrated feature of the life history of the species (Chapter 6). More particularly, the species judged on the basis of other features (the presence of autogamy, the response to inbreeding, the mode of mating type determination, etc.), to be inbreeders, generally have a short life span; species interpreted as outbreeders tend to have a long life cycle—consistent with their need to journey among strangers before mating. While the length of life might be the primary determinant of all other special features of the natural history of the species, it seems more likely that the length of life has been adjusted along with the other characteristics into an integrated life style. In any case, and without dispute, evolutionary modifications must have occurred in the life span, and these modifications establish the modifiability of the mechanisms of somatic senescence.

Smith-Sonneborn has recently reported that somatic senescence in Paramecium may be delayed by certain experimental manipulations. Particularly, when low doses of ultraviolet light are followed by visible light, clones are able to undergo additional divisions to 50% or more beyond the controls. These observations suggest that an increase of life may be achieved by increasing the activity of the DNA repair enzymes. By extension, the evolutionary adjustment of life span might reflect changes in the efficiency of repair systems.

B. Germinal Senescence

The loss of the ability to produce viable progeny in conjugation is an age correlated phenomenon, but it differs from the changes associated with somatic senescence at least quantitatively. The loss of fertility generally occurs much less regularly. It may occur in some lines at a relatively young age, before they begin to show loss of somatic vigor. On the other hand, some vigorous progeny may be produced by clones well into their senescent decline.

Perhaps the most generally satisfactory interpretation of germinal aging is that it involves genetic accidents—gene mutations and chromosome aberrations. In a species with somatic senescence, sexual fertility may be lost at any time by the occurrence of micronuclear mutations. Because the micronuclear genotype is not expressed, these mutations continue to be carried and accumulate with time. Eventually all micronuclei will contain one or more dominant lethals and their capacity to develop normal macronuclei will be greatly reduced or lost entirely. This constant accumulation of mutational damage does not entirely explain germinal aging, however. The rate of germinal aging in Paramecium probably increases sharply at the time of onset of somatic senescence. Certainly "young" micronuclei introduced into "aged" cytoplasm respond with aberrant behavior and chromosome alterations. The error control systems of the aged cell become deficient and even normal nuclei in a deficient environment are unable to maintain their integrity. In this sense the terminal stages of germinal aging may be a secondary consequence of somatic senescence. Organisms such as some of the Tetrahymena species that lack a phase of somatic senescence may also lack the terminal *acceleration* phase of germinal aging, even though they manifest the random time invariant accumulation of micronuclear damage.

C. Temporal Programs

The somatic senescence of ciliates is an example of an *intrinsic temporal program.* At about the same time in all its sublines a clone begins its vegetative decline. This decline does not depend on a changed external environment, but occurs under essentially steady state conditions, as in a chemostat, or with daily reisolation into fresh medium.

Ciliates are also subject to *extrinsic programming,* and this kind of response is sometimes confused—at least semantically— with intrinsic programming. The normal laboratory growth cycle begins with the inoculation of a fixed amount of culture medium with one or more cells. If the inoculum is starved, a lag period is observed before the beginning of cell division. The lag phase is followed by a fast exponential phase (or *infradian phase*) of rapid growth. As the nutrients are exhausted and as toxic metabolites increase, the division rate declines and the culture enters the *circadian phase,* so called because the infrequent cell divisions are coupled to the cellular clock and may be entrained to synchrony. In a sense a culture in early exponential growth is "young," and a culture in circadian growth is "old," but these changes in cellular composition and behavior are extrinsically controlled by patterned changes in the environment. The cells modify their surroundings and are in turn modified by them. Possibly the changes that occur in a closed culture vessel of ciliates are analogous to the changes that occur in the closed corporeal system of a higher vertebrate. Perhaps aging in a metazoan is caused by the progressive toxification of a cellular environment that cannot be chemostatically maintained.

Regardless of the aptness of the analogy between a starved culture and an old metazoan, a young ciliate in terms of the cultural growth cycle need not be a young ciliate in its clonal life history. The clonal age of a culture is usually measured by the number of cell divisions that have occurred since fertilization, rather than the number of cell division since inoculation. A paramecium clone 200 cell divisions old does not behave the same way as a clone 100 cell divisions old, even if they are inoculated simultaneously into identical culture vessels.

The problem we need to consider here is the mechanism whereby long term intrinsically regulated changes are directed. Perhaps the most significant constraint on explanations of clonal aging is the time over which the regular modifications occur. Vegetative decline in *P. tetraurelia* has its onset after about 50 cell divisions. Other specific life history events in different ciliates may have their onset after 20, 100, or 400 cell divisions. Some of these are long term programs by any measure, particularly when we consider that only about 50 cell divisions intervene between the egg and the adult in humans.

The length of the life history program is important because it excludes the simplest and perhaps the most attractive explanation for temporal patterns, that is, dilution. We might imagine, for example, that some biological process is inhibited by some material accumulated during conjugation. This material is diluted by cell division until cells are produced with no inhibitor molecules and the previously inhibited process is allowed to occur. The length of time the inhibition would persist would depend on the number of molecules originally stored. A large store of inhibitor molecules would result in a relatively long delay in expression.

On what time scale is such a model of temporal programming effective? What is a reasonable limit to the number of inhibitor molecules that might be stored for such purposes? The absolute upper limit of inhibitor molecules is the absolute number of molecules contained in a cell. A large cell like Paramecium, of course, has more molecules than a small cell like a yeast. Nevertheless, the total number of molecules is probably not greater than 10^{13}. If every molecule is an inhibitor molecule, and if a single molecule is an effective inhibitor, and if the dilution process is so exact as to prevent any clustering of inhibitor molecules in pedigrees, how many cell divisions are required to generate uninhibited cells? The number of divisions (N) is calculated by $2^N = 10^{13}$, for after N divisions the number of cells is equal to the number of molecules originally present. One additional division will yield a population of cells of which at least half are uninhibited. N so calculated is equal to about 45–50 cell divisions. Even under these most stringent restrictions, temporal programs are limited to an extent of only about 50 cell divisions. Yet the programs we observe in ciliates may be effective over two or four or eight times such *exponential* intervals. Dilution programming is suitable only for short term effects.

The simple dilution model may, of course, be modified by converting it into a *differential production* model; that is, instead of postulating the synthesis of inhibitor only during conjugation, we might imagine that the inhibitor is also produced during vegetative growth but at a rate somewhat lower than that of the total cell substance. A mirror image model can be developed by postulating an inducer instead of an inhibitor and considering the gradual accumulation rather than the gradual dilution of the controlling molecules. Both models may require threshold action instead of single molecule control.

Models of differential dilution and differential accumulation are superficially attractive, but are difficult to apply precisely. They require exquisitely fine controls on molecular synthesis, and if a random distribution of controlling molecules to daughters is proposed, they are likely to yield great diversities in times of threshold achievement in different sublines. A major consideration in evaluating the plausibility of such models, therefore, is the actual precision of the temporal controls: how much variability is observed in the time at which specific biological events occur in independent lineages of the same genetic constitution? Relatively little published information is available on such topics, particularly for the long term events. Long term events by their very nature require a long period of observation, and the replication of observations on an adequate scale requires that equipment and space be tied up for extended periods.

However, senescent decline is not the only programmed event in the life history and some of the other events are more economically analyzed. In *Euplotes crassus* the phenomenon of *senescent selfing* (see Chapter 6) occurs in clones heterozygous for mating type alleles as they approach the period of decline. This phenomenon, however, is not much more accessible for analysis than the decline in growth rate itself, for it requires some 400 cell generations and many months. Other life cycle events may be more useful. The timing of sexual maturation has been subjected to some analysis. Different strains of the same species of Tetrahymena, for example, mature at different times, and the hybrids have intermediate timing. Mutations that greatly reduce the immaturity period have been described. For Paramecium some data indicate that the clonal age of the parent may affect the immaturity interval in the progeny.

Thus far the ciliate studies on temporal programming are significant primarily because they show that the controls are in fact intrinsic, that they may affect events over long periods of time, and that in at least some cases they are remarkably precise.

D. The Necessity of Senescence

Sonneborn has recently observed that senescence, in the sense of a programmed termination to the life cycle, is not to be found among organisms without a significant diploid phase, while diploids—even diploid microbes—often undergo "planned" ·obsolescence. This observation, combined with the considerations developed in Chapter 6, leads to a partial rationalization of the significance of senescence.

A haploid organism, we argue above, has an ecogenetic strategy of adjusting to environmental hazards by exploiting mutational variety. A haploid organism constantly exposes its genetic diversities and maintains no reserves. The diploid organism, in contrast, insofar as allelic dominance applies, has the capacity to maintain a reservoir of cryptic variability. The degree to which the reservoir is maintained is an adjustable ecogenetic parameter. A diploid that undergoes frequent selfing, or mating with close relatives, constantly opens its mutational reservoir to environmental inspection; it has a fundamentally "haploid economy."

The outbreeding economy, as we see above, involves a systematic avoidance of mating with close relatives, and hence an avoidance of homozygosity. Outbreeders capitalize, in an imperfectly understood way, on recombinational variety arising from mating with strangers. They do not avoid mutations, however, and in fact require mutations as the raw materials for the recombinational lottery. In any case, their deferred sexual gratification leads to an increase in the mutational reservoir with advancing clonal age.

The consequences of an accumulating mutational load may be

described in general terms. Most of the mutations, though of possible utility under some circumstances, are likely to be deleterious. Perhaps the outbreeding strategy is based on a greater likelihood of useful variants from recombinational than from mutational sources. An aging organism, however, has a greater and greater probability of exposing in its progeny its deleterious mutations. These harmful genetic changes are only partially shielded from exposure in a simple diploid, because perfect dominance is unusual. However, organisms with a segregated germplasm (such as ciliates and metazoa) accumulate even dominant mutations without limit, because the affected genes are not expressed in somatic functions; but once such mutations are delivered to the sexual progeny, and expressed in somatic structures, their harmfulness becomes apparent.

Beyond a certain time of accumulation, therefore, the cryptic reservoir of the outbreeder becomes disadvantageous. The technique of sterilizing the germplasm as a means of decimating a population is a familiar, even a commercially exploitable, device. One might expect, therefore, that a natural method would be devised for disposing of mutation-laden old outbreeders without paying all the costs of the mutational load. Senescence could then be considered a method for disposing of individuals when they become hazardous to the species.

The difficulty with this explanation is that it specifies the advantages to the species and not to the individual manifesting an aging program. It requires group selection to operate, and group selection encounters theoretical objections. These objections should not be the final word. Teleological interpretations can sometimes be rephrased in teleonomic terms, and advantages to the species may be linked in some as yet obscure way to an improvement in the fitness of the individual. Without question the life histories of organisms, including their senescence and death, are shaped by natural selection, even if we have not yet identified the mode of leverage of the selective forces.

SUMMARY

Many ciliates have finite life spans with well-defined intervals of immaturity, maturity, and senescence. The life spans may vary among different species from a few weeks to several years. Within a species, however (or more precisely, within a genotype within a species), the events may occur with great precision after long intervals of cell division. The number of cell divisions is associated firmly with the time measuring system in most ciliates, but the mechanism of measurement is not known.

The life history of a ciliate may be viewed as an intrinsic genetic program, relatively immune to transient environmental conditions. The capacity to employ intrinsic genetic programs may be a general eukaryotic capability, as fundamental as the other eukaryotic temporal device—the circadian clock. Whether the mechanism of temporal programming in ciliates is in fact similar to that underlying the behavior of explanted vertebrate cells remains to be determined.

RECOMMENDED READING

Bleyman, L. L. 1971. Temporal patterns in the ciliated protozoa. In *Developmental Aspects of the Cell Cycle* (T. L. Cameron, B. M. Padilla, and A. M. Zimmerman, Eds.), Academic, pp. 67–91.

Ehret, C. F. 1974. The sense of time: Evidence for its molecular basis in the eucaryotic gene-action system. *Adv. Biol. Med. Phys.,* **15,** 47–77.

Fukushima, S. 1975. The clonal age and the proportion of defective progeny after autogamy in *Paramecium aurelia. Genetics,* **79,** 377–381.

Heckmann, K. 1967. Age-dependent intraclonal conjugation in *Euplotes crassus. J. Exp. Zool.,* **165,** 269–277.

Jennings, H. S. 1944. *Paramecium bursaria:* Life history. I. Immaturity, maturity and age. *Biol. Bull.,* **86,** 131–145.

Miwa, I., H. Nobuyuki, and K. Hiwatashi. 1975. Immaturity substances:

material basis for immaturity in *Paramecium. J. Cell Sci.,* **19,** 369–378.

Nanney, D. L. 1974. Aging and long-term temporal regulation in ciliated protozoa. A critical review. *Mech. Ageing Dev.,* **3,** 81–105.

Siegel, R. W. 1967. Genetics of aging and the life cycle in ciliates. *Symp. Soc. Exp. Biol.,* **21,** 127–148.

Smith-Sonneborn, J. and S. R. Rodermel. 1976. Loss of endocytic capacity in aging Paramecium: the importance of cytoplasmic organelles. *J. Cell Biol.,* **71,** 575–588.

Smith-Sonneborn, J. 1979. DNA repair and longevity assurance in *Paramecium tetraurelia. Science,* **203,** 1115–1117.

Sonneborn, T. M. 1954. The relation of autogamy to senescence and rejuvenescence in *Paramecium aurelia. J. Protozool.,* **1,** 38–53.

Sonneborn, T. M. 1978. The origin, evolution, nature and causes of aging. In *The Biology of Aging* (J. A. Behnke, C. E. Finch, and G. B. Gairdner, Eds.) Plenum, pp. 361–374.

Corticogenes and Structural Inertia

10

BIOLOGISTS FIRST ENCOUNTERING THE CILIATES ARE likely to be impressed with the remarkable complexity of their cortical organization. These acellular (or unicellular) creatures manifest a degree of surface specialization unexpected in reputedly simple organisms. We have been conditioned to think in terms of the developmental systems of metaorganisms, in which we know that cellular differentiation is a key to structural and functional specialization. The ciliates are organisms without cells, but they are nevertheless capable of differentiating some of their isolated parts to perform special tasks.

One possible explanation for the ciliates' regional specialization is a cryptic pseudocellularization. Instead of functionally differentiated cells, local reservoirs of a few genetic elements might account for special structures or functions. These proposed cytoplasmic reservoirs were supposed to be populated by genetic elements called *cytogenes, plasmagenes,* or *corticogenes.* The postulated genetic substations could account in principle for diversities of structure and function in an uncompartmentalized protoplasm. Corticogenes could also account for a particular morphogenetic mystery: the fact that new ciliate structures generally arise in close association with preexisting structures of the same kind. If genetic determinants are localized in the regions of the structures they specify, the local replication of those determinants would be expected to be coupled with the replication of the specialized structures.

This hypothesis of the corticogene led to two kinds of exploration: a search for chemical evidence of DNA in the cortex, and a search for evidence of genetic distinctions among cytoplasmic regions. We consider first the studies on cytoplasmic specificity and continuity and return later to the problem of cortical DNA.

A. Microsurgical Studies on Stentor

One of the earliest and most decisive tests of the hypothesis of genetic specialization of cytoplasmic regions came from microsur-

153

gical studies on the giant ciliate Stentor. The cylindrical structure of the primitive ciliate is modified in Stentor into a cornucopial funnel. The membranelles at its large end sweep food particles into its oral apparatus. Although Stentor appears at first glance to be radially symmetrical, a more careful examination shows an asymmetry about its longitudinal axis. The ciliary rows, running from anterior to posterior, are not equally spaced. In one region they are very close together, and as one moves around the cell to the right the rows become more and more widely spaced (Fig. 10–1). This spacing of the ciliary rows is readily visible because of regions of pigmented cortex between the rows. The pigmented stripes become wider and wider around the cell, until one returns to the starting point and the region of narrow stripes. The region of

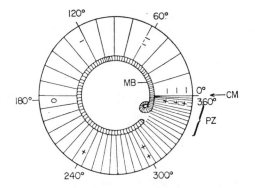

Fig. 10–1. A diagram of Uhlig's "circular gradient" in *Stentor coeruleus.* The cell is seen from above, looking down on the anterior end, with the membranellar band contracted so that the stripe organization on the body surface posterior to it can also be seen. The gradual decrease in width of the stripes from 0 to 360° reflects a corresponding increase in gradient values from − − − to + + +. The contrast meridian (CM) is the site of the greatest difference in gradient values; the zone in which the primordium develops (PZ) is on the + + + side of this gradient discontinuity. Modified slightly from Abb. 6 of Uhlig (1960) by Frankel, with permission of the author. Reprinted with permission from Frankel, J. 1974. Positional information in unicellular organisms. *J. Theor. Biol.* **47,** 439–481. Fig. 7, p. 459. Copyright by Academic Press Inc. (London) Ltd.

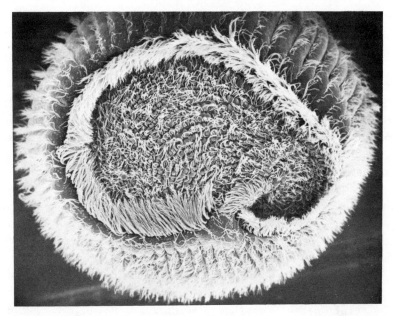

Plate IV. Scanning electronmicrograph of *Stentor coeruleus.* Oral view of a contracted specimen. Magnification: 500×. Courtesy of Jerome J. Paulin.

abutment between narrow stripes and broad stripes is a striking discontinuity in the cortex, and it has major morphogenetic significance.

The basal bodies for new oral membranelles arise from a few ciliary rows with narrow stripes in the region of contact with the wide stripes. When the Stentor divides, both of its daughters receive segments of the stomatogenic rows, and in recurring generations the oral membranelles continue to be derived from descendants of these same ciliary rows.

The corticogene hypothesis would propose that the basal bodies of the stomatogenic rows are genetically distinct from those in other ciliary rows. The special genetic elements associated with these basal bodies would explain their ability to give rise to oral membranelles. The direct continuity of stomatogenic rows from generation to generation would assure the perpetuation of the genetic distinction of these cortical elements.

Stentor cannot be grown on a defined medium, its sex life cannot be controlled, and little is known about its physiology. But Stentor is very large, some species ranging up to 500 micrometers or more, and it is remarkably tolerant of surgical mutilation. Tartar, Uhlig, and others have exploited these special characteristics to obtain a test of the corticogenic interpretation. If the stomatogenic rows contain genetically distinct corticogenes, the surgical removal of the rows should lead to a failure of oral morphogenesis; but it does not. The activities previously associated with these rows are taken over by the next rows on the right. A ciliary row behaves as a stomatogenic row not because of its unique constitution, but because of its unique position.

This interpretation of morphogenetic activity in terms of position rather than of constitution is buttressed by many observations and experiments. The structural discontinuity in the circumcellular gradient of stripe widths is apparently a key to stomatogenic activity. Wherever regions of stripe contrast are found, in a variety of surgical reconstructions, stomatogenic function follows.

One might perhaps argue still for the existence of genetic substations, but if they do exist their corticogenes would have very different significance and would be of little explanatory value. One might, for example, propose that various classes of genetic information (DNA or RNA) are sequestered by specific cytoplasmic regions (such as the stomatogenic rows) and are responsible for regional specificity of synthesis and incorporation; but the genetic information would have to be only sequestered by the distinctive regions and would not find its unique source in the regions. The functional differences among cytoplasmic regions would be responsible for their diverse genetic content, rather than the reverse. The transfer of functions to adjacent regions, when cortical areas are removed, or the emergence of functions in regions of artificial discontinuity, are not explained by the genetic components of the regions even if the genetic distinctions were to develop in the new areas. Rather, the genetic differentiation would be a consequence of a prior differentiation of function, based on spatial information.

B. Cortical Slippage in Tetrahymena

The principle of cortical totipotency is not limited to Stentor, but has been verified wherever suitable tests could be devised throughout the ciliates. An example of such tests is provided by a morphogenetic variant in Tetrahymena. The new oral apparatus in these species arises as a field of basal bodies localized to the left of the stomatogenic ciliary row (row 1, Fig. 10-2). The basal bodies become organized into the four membranellar primordia and give rise to the cilia making up the *undulating membrane* (UM) and the *adoral zone of membranelles* (AZM) (Fig. 10-3). When the cell divides, the old oral apparatus goes to the anterior daughter, and the new oral structures go with the posterior cell. The stomatogenic row, however, is divided, with half going to the anterior and half to the posterior daughter. These portions of the original stomatogenic row elongate during the next cell cycle, by the addition of new basal bodies anterior to preexisting basal bodies; at the appropriate time in the cell cycle, basal bodies are

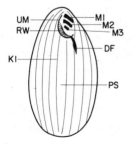

Fig. 10-2. Diagram of the cortical structure of Tetrahymena. (DF) deep fiber; (K1) kinety (ciliary row) 1; (M1, M2, M3) membranelles 1, 2, 3; (UM) undulating membrane; (RW) ribbed wall; (PS) site of the oral primordium. Reprinted, by permission, from Frankel, J. and N. Williams 1973. In *Biology of Tetrahymena* (A. M. Elliott, Ed.), Dowden, Hutchinson and Ross. Fig. 1, p. 377.

again proliferated into the space to the left of the row and form the primordium of a new oral apparatus. The stomatogenic rows within a clone of cells may be thought of as a continually elongating chain of basal bodies, periodically segmented (at cell division) and periodically induced into oral morphogenesis. The physical continuity and persistence of function of this ciliary row are consistent with its genetic individuality.

A variant strain of a related species challenges such an interpretation, however (Fig. 10–4). In this strain about half the cells undergo stomatogenesis in the normal way. In the other half, the ciliary row to the left of row 1 (row *n*) begins to generate a morphogenetic field, either exclusively or in addition to row 1, and gives rise to an oral apparatus for the posterior daughter. The posterior segment of row 1 in these cases becomes not row 1 in the posterior daughter, but row 2. In the next morphogenetic cycle of the posterior daughter, stomatogenesis again has a 50%

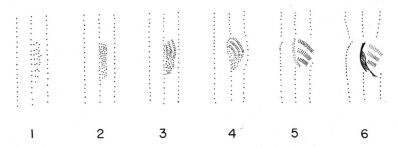

Fig. 10-3. Sequence of events during development of the oral primordium in Tetrahymena. Each diagram represents the midregion of ciliary row 1 (K1) plus the two adjacent ciliary rows. The oral primordium is represented as it appears at six arbitrarily designated stages of development. In stage 0 the primordium has not begun to develop, but it appears as a small cluster of basal bodies at stage 1. In stage 2 it is represented as a field of basal bodies that increases in size and organization until the time of cell division at the end of stage 6. Reprinted, by permission, from Frankel, J. and N. Williams, 1973. In *Biology of Tetrahymena* (A. M. Elliott, Ed.), Dowden, Hutchinson and Ross. Fig. 2, p. 377.

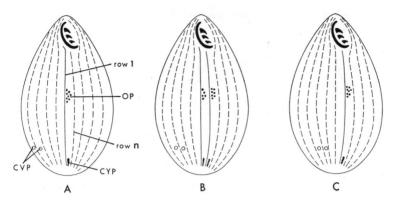

Fig. 10-4. Representation of "cortical slippage" in a variant strain of Tetrahymena. (*A*) The normal situation in which the oral primordium (OP) develops on the (cell's) right postoral ciliary row. (*B*) The rare intermediate situation in which oral primordia begin developing on both postorals. (*C*) The situation in which the oral primordium has moved from row 1 to row *n* (the left postoral row) with a simultaneous shift of the cytoproct (CYP) and the contractile vacuole pores (CVP) one row to the (cell's) left. Reprinted, by permission, from Nanney, D. L. 1967, Cortical slippage in Tetrahymena. *J. Exp. Zool.*, **166**, 163-169. Fig. 1, p. 164.

probability of slipping to the left. In these cases the original row 1, which became row 2 in the "slipped" daughter, becomes row 3 in the slipped granddaughter. Since the total number of rows is usually about 20, and since stomatogenic slippage occurs about half the time, in the course of 40 cell divisions, stomatogenesis will usually have slipped to the left, row by row, all the way around the cell and will return to its original position. All the morphogenetic activities capable of occurring in any part of the cell will eventually occur in association with lineal descendants of each and every ciliary row.

The different ciliary rows are thus demonstrably equivalent in morphogenetic capacity. Their realized capabilities depend, however, on their position both in space and time. The corticogene hypothesis provides no explanation for the morphogenetic capacities of cytoplasmic regions.

C. Doublet Inheritance in Paramecium

The regional specialization of the ciliate cortex is related to the cortical variation that arises within a clone. Cell parts seem to have hereditary continuity and certain cell types persist indefinitely. The persistence of different structural characteristics is not, however, associated with different genetic determinants. One of the best studied and simplest examples of hereditary cortical differences is that of doublets in *P. tetraurelia.*

Ordinary, or *singlet,* paramecia have a single, but complex, oral apparatus (Figs. 3–1 and 10–5) on their ventral surface. They also have a *cytoproct* on the ventral surface, posterior to the mouth, and two *contractile vacuole pores* (CVPs) on the dorsal surface. Occasional *doublet* cells arise, either spontaneously or by specific induction. These doublets are about twice the normal size and

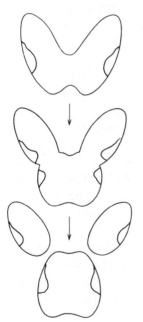

Fig. 10–5. The origin of doublet paramecia following delayed separation of conjugants. In some instances normal singlets bud from the anterior end of the fused pair but leave the united posterior portions as a stable doublet. Reprinted, by permission, from Grell, K. G. 1973, *Protozoology,* Springer-Verlag. Fig. 240a, p. 274.

have two ventral surfaces and two dorsal surfaces. Their important feature in the present context is their stability. Generally singlets give rise to singlets and doublets give rise to doublets.

Sonneborn carried out exhaustive analyses of the physical basis for the hereditary differences between singlets and doublets. Doublets can be crossed to singlets and the basis for their differences can be evaluated by standard genetic procedures. Particularly (Chapter 7), crosses can be made in this species under conditions of controlled cytoplasmic fusion. Conjugation may occur under circumstances in which only nuclei are exchanged, and nuclear equivalence is established in the originally singlet and doublet mates. In such pairs the exconjugants with identical genetic markers nevertheless give rise, respectively, to singlet and doublet clones.

Crosses may also be arranged to permit extensive cytoplasmic exchange, so that conjugants not only come to have equivalent nuclear genes, but also share cytoplasmic symbionts or different mitochondrial mutations (Chapter 15). When such crosses are made, the results with respect to cell form are the same; the doublet exconjugant gives rise to a doublet clone and the singlet exconjugant produces a singlet clone (Fig. 10-6). The "determinants" for singularity as opposed to duplicity are not mixed with the fluid cytoplasmic elements, but remain associated with the relatively rigid cortex.

We cannot adequately consider the mechanism of doublet maintenance without some attention to the origins of doublets. If all we knew about the doublets was that they arise infrequently, are perpetuated with fidelity, and do not differ from singlets in nucleus or cytoplasm, we might think that they depend on mutant corticogenes. In the mutant form corticogenes would change the local patterns of synthesis and/or association of gene products and yield a novel cellular structure; but doublets may be produced at will in high frequency by treatment of conjugating cells with immobilizing antiserum (Chapter 12) prior to the time of normal

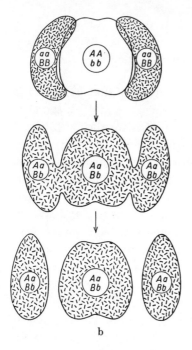

Fig. 10-6. Genetic consequences of a cross with cytoplasmic exchange between a doublet paramecium and two singlets. Nuclear exchange is monitored by the phenotypes controlled by nuclear genes (*A, B*). Cytoplasmic exchange is followed by means of the killer trait, controlled by the symbiotic bacterium kappa (Chapter 15). Reprinted, by permission, from Grell, K. G. 1973. *Protozoology,* Springer-Verlag. Fig. 240b, p. 274.

b

separation. At appropriate concentrations of antiserum, many pairs miss the magic moment of separation and develop broad areas of cytoplasmic continuity. They eventually restructure their cortical components into a symmetrical double configuration in which approximately normal morphogenesis and fission can continue indefinitely.

It is hard to believe that the antiserum specifically mutates the putative corticogenes, all or even many of them, so that they prescribe a new cellular architecture. Simple exposure of exconjugants to antiserum has no such effect—even if the application occurs a few minutes after separation. Moreover, when morphogenetic accidents yield singlets from doublets, the singlets remain as singlets and do not revert to the doublet state. No evidence is available for a separation of the determinants of the cell form as distinguished from the manifested form itself.

Doublets are also easily produced in other ciliates, and some of these are very stable. Again the circumstances of origin are informative. Doublets in Tetrahymena may be produced, as in Paramecium, by interfering with separation of conjugants. Fauré-Fremiet induced doublets in Tetrahymena by interfering with cell separation at fission; low concentrations of formaldehyde applied at the critical time in the cell cycle led to fusion of the daughter cell bodies, a reorganization, and, often, a stabilization of a symmetrical and persistent duplex structure.

The most plausible interpretation of these phenomena lies in the concept of *structural inertia*. The organization of the cortical components conditions the placement of new cortical elements. The cortex is capable of stable (or metastable) reproduction in any of several patterns of organization. The pattern of organization may be modified by interference at critical stages of morphogenesis, but tends to perpetuate itself under usual conditions.

The cellular architecture is by no means independent of genic control. But these genes need not be localized in the cortex. Nuclear genes specify the cortical molecules and in so doing place limits on the patterns of association of the molecules and of cortical organelles. The genes allow a degree of freedom in the deployment of the molecules and of their combinations. Preexisting patterns of organelles influence the placement of new structures and provide a kind of cellular memory of historical events. Structural inertia permits the coexistence of two or more biotypes with the same genetic components and same molecular constitution. Molecular composition does not uniquely prescribe cellular architecture.

D. Inverted Ciliary Rows in Paramecium and Tetrahymena

The most convincing evidence concerning the concept of structural inertia (*cytotaxis* of Sonneborn or *structural guidance* of Frankel) comes from an analysis of organizational variants in which ciliary rows are reversed in polarity. We mention earlier (Chapter 2) that each ciliary unit is a complex asymmetric structure, with a well-defined inside and outside aspect, a left and a right margin, and an anterior and posterior face. Ciliary units are organized characteristically into rows of units of corresponding symmetry, and they reproduce that symmetry by the dynamics of their growth. New basal bodies arise anterior to preexisting basal bodies as cylinders of microtubules directed at right angles to those of the "parent" basal body. The new basal body elongates, moves away from its origin, tips toward the cell surface, and eventually develops a ciliary shaft and the other fibrous and membranous components needed to make it a complete ciliary unit in its own right.

Paramecia with one or more reversed ciliary rows have been experimentally developed by Beisson and Sonneborn, using as their basic material conjugating cells whose separation has been prevented by antiserum treatment (Fig. 10–7). We note earlier that doublets can be produced with this technique when the antiserum concentration and the time of application are appropriate.

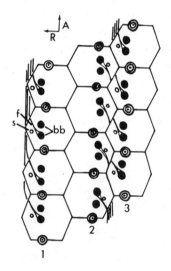

Fig. 10-7. Diagram of an inverted ciliary row in the cortex of Paramecium. Rows 1 and 3 have the normal orientation of the hexagonal ciliary units. The basal bodies (bb) are slightly off-center to the cell's right (R), and the parasomal sac (s) is on right margin of the ciliary unit. The kinetodesmal fiber (f) moves anteriorly and to the right from the posterior basal body. All these relations are reversed in the inverted row 2. Reprinted, by permission, from Sonneborn, T. M. 1970. Gene action in development. *Proc. Roy. Soc. Lond. B.,* **176,** 347–356. Fig. 13, p. 353.

In other cases cytoplasmic bridges of various width link the conjugants, and the cells often become oriented in a heteropolar configuration as they begin to pull apart. The eventual separation is never as clean as separation at the normal time, and parts of the cortex from one cell may adhere to the surface of the mate, sometimes in reversed polarity.

The subsequent histories of these patched cells have been followed carefully. Cells with reversed patches near the midline are particularly interesting, because ciliary rows in Paramecium elongate primarily by the production of basal bodies in the midregions (Fig. 10–8). A patched cell with a segment of inverted cortex in the midregion will eventually produce some offspring in which the patch extends from end to end. If isolated these cells give rise to pure cultures with inverted stripes that breed true for hundreds of cell generations. They can be distinguished by cytological methods, but also have behavioral peculiarities.

As in the case of the doublets, genetic analyses have been carried out to localize the determinants of the hereditary differences between normal and inverted cells. Again, no nuclear or fluid cytoplasmic influence can be demonstrated. The inheritance of ciliary row polarity is determined by the cortex of the cell.

Given the circumstances of origin of the reversed ciliary rows, it is hard to believe that the reversed rows contain different genetic elements, or indeed molecules of any kind differing from those in adjacent rows of normal polarity. Two patterns of organization may coexist in the same cell—in adjacent ciliary rows—and two patterns of organization may coexist in the same culture—as in cells with and without reversed rows—without any implication of structural genic differences, functional genic differences, or indeed molecular differences of any kind.

Cells with inverted ciliary rows have also been constructed by Ng and Frankel in *T. thermophila,* but by a different procedure. Frankel selected a number of conditional mutations capable of normal fission at 30 °C, but incapable of completing the process at high temperatures (Chapter 16). Certain mutants exposed to high temperatures during fission remain as tandem duplex cells, which undergo irregular reorganization processes, often leading to heteropolar duplexes. After return to lower temperatures, these duplexes tend to produce singlets as they resume division, and some of these contain segments of inverted cortex (Fig. 10–9). Even one inverted row results in modified swimming behavior so that living cells with reversed rows may be selected and cloned. Again such cultures have been studied, and the structural altera-

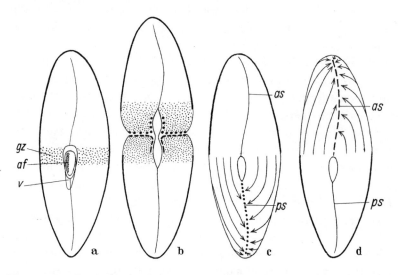

Fig. 10–8. Cortical events associated with cell division in Paramecium. (gz) Growth zone, the region in which most new basal bodies arise. (v) Vestibulum, the sunken area of the oral apparatus leading to the gullet. (af) Anlagen field, the special field of basal bodies on the wall of the vestibulum that gives rise to the oral apparatus of the posterior daughter. After fission the ciliary rows of the anterior cell move posteriorly to meet in a new posterior junction (ps). The ciliary rows of the posterior cell in a similar way move forward to constitute the anterior junction (as). A misoriented segment of cortex in the midregion is differentially replicated and extended eventually to both ends of most cells. Reprinted, by permission, from Grell, K. G. 1973. *Protozoology,* Springer-Verlag. Fig. 122, p. 134. After Schwartz, V. *Naturwiss.* **50,** 631–640, 1963.

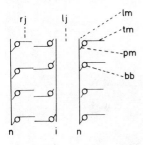

Fig. 10–9. Representation of an inverted ciliary row in Tetrahymena. In the normally oriented rows (n) the basal bodies (bb) and their associated cilia are close to the longitudinal microtubular bands (lm) on their right. Two other basal body associated fibrous components are also recognized, the transverse microtubular bands (tm) directed to the cell's left and the postciliary microtubular bands (pm) directed posteriorly to the right. Again, as in Paramecium, references to left and right refer to the organism and *not* to the viewer. The junctions between normal and inverted ciliary rows are distinctive; particularly, the right junction (rj) of an inverted row has two transverse connectives, and the left junction (lj) has none. Reprinted from Ng, S. and J. Frankel 1977. 180° rotation of ciliary rows and its morphogenetic implications in *Tetrahymena pyriformis. Proc. Natl. Acad. Sci. U.S.* **14,** 1115–1119. Fig. 5b, p. 1117.

tion has been shown to be "hereditary." The implication of the Tetrahymena variants is identical to that of the Paramecium variants: novel patterns of organization may be perpetuated indefinitely in the absence of distinctive genetic elements, by virtue of the morphogenetic role of that structure. Organisms use preexisting structure as scaffolding in their growth processes, and differences in preexisting structure may be perpetuated indefinitely through a process of structural inertia.

One should be wary of concluding that microscopically visible structures are the operative structures in the perpetuation of structural information. The visualized structures may be indicative only of invisible structures responsible for morphogenetic patterns. An important illustration of this possibility comes from Grimes' study of encystment in the hypotrich ciliate *Oxytricha fallax.* These organisms may occur as stable singlets or doublets,

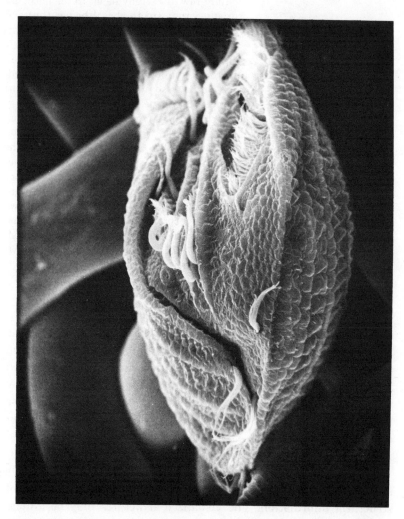

Plate V. Scanning electronmicrograph of *Euplotes aediculatus* Homopolar doublet with two complete sets of cortical structures. Magnification: 840✕. Courtesy of Gary W. Grimes.

168

much as in the case of Paramecium or Tetrahymena. Unlike Paramecium (and most Tetrahymena) Oxytricha encysts under certain conditions. While in the cyst state all traces of the cortical units disappear; serial sections of entire cysts reveal no basal bodies. Yet, when the cysts are activated and vegetative cells emerge, doublet cells are produced from the cysts of doublets, and singlets emerge from the singlet cysts. The pattern of organization persists through the dissolution of the structures that make it evident. The controlling molecular substructure may be disrupted, however, by physical agents—which may destroy the doubled state and convert a doublet into a singlet.

E. Incomplete Doublets in *Oxytricha fallax*

One final example of a stable cortical variant comes from Grimes' study of laser microbeam effects on Oxytricha doublets. A clone with a new morphotype developed from a treated doublet; the cells were neither normal singlets nor complete doublets, but had a peculiar "humped" appearance (Fig. 10–10). The "hump" was due to the presence of an extra row of marginal cirri that persisted after most of the duplicated cortical structures were destroyed. The new basal bodies in the marginal cirri (unlike those for many other cirral structures) regularly arise in association with the old basal bodies. In the humped cells the supernumerary cirral row developed new basal bodies and elongated at the usual time for cortical morphogenesis; the extra row was divided between the two daughter cells. The extra rows in the posterior daughters were stable generation after generation. Those in the anterior daughter sometimes persisted, but more commonly became aberrant and were resorbed.

One may scarcely doubt that the presence of a row of marginal cirri plays a role in the perpetuation of this structure. Oxytricha can maintain during vegetative replication at least three reasonably sta-

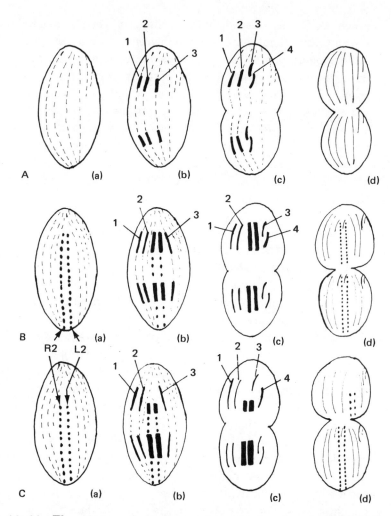

Fig. 10-10. The patterns of morphogenesis on the dorsal surfaces of (*A*) normal singlets and (*B* and *C*) "humped cells" of *Oxytricha fallax*. Because of difficulties in maintaining the conventional internal circumcellular perspective for dorsal structures, we refer here to structures on the reader's left and right. (*A*) Prefission dorsal morphogenesis in a normal singlet. (*a*) The pattern of ciliary (or bristle) rows in a nondividing cell; the ciliary rows are represented by dashed lines. Four rows extend from end to end, but two short rows are restricted to the anterior

ble cortical configurations—the singlet, the doublet, and the humped cell, in the absence of any genetic distinctions. The most interesting feature of the humped cell analysis, however, derives from observations on cysts. Unlike the full doublet condition, the humped (partial) doublets do not persist through encystment, but revert to singlets. This behavior suggests at least a duality of devices in the maintenance of structural assemblies—the local constraint of morphogenetic activities by cellular scaffolds and the "field" constraints defining the total pattern. A field can apparently persist without elaborate organellar manifestation, but local constraints at variance with normal field expression may be lost when their organellar base is removed.

F. The Question of Cortical DNA

The evidence summarized above argues that local structures and functions in a complex ciliate cannot be *explained* by the presence

←———————————————————————

(reader's) right side. The primorida for the four long rows (*1–4, b, c*) arise within the three left rows and are represented as solid dark lines. A late constriction stage (*d*) shows all the new rows (solid lines) complete in both anterior and posterior daughters. The short rows in both daughters derive from the primordium of the marginal cirral row on the ventral surface. (*B*) Morphogenesis in a humped cell that has a pair of additional cirral rows (R2, L2) in the middle of the dorsal surface. The new dorsal ciliary primordia (*1–4, b, c*) arise on either side of the primordia for the extra marginal cirri. A late constricting cell (*d*) has the aberrant pattern replicated in both daughters. The short anterior rows for the left part of the dorsal surface derive from one of the dorsal cirral rows (R2), while those for the right part come as usual from the marginal cirral row (R1) on the ventral surface. (*C*) Alternative morphogenetic pattern in humped variants in which the cirral rows do not extend so far anterior (*a*). The primorida for the cirral rows in the anterior cell are short (*b, c*) and fail to give rise to the short ciliary rows. As a consequence the anterior daughter fails to maintain the humped pattern in its complete form, but the posterior daughter is without defect (*d*). Reprinted, by permission, from Grimes, G. W. 1976. Laser microbeam induction of incomplete doublets of *Oxytricha fallax. Genet. Res. Camb.*, **27**, 213–226. Fig. 9, p. 218.

of distinctive genetic elements in different cytoplasmic compartments. The evidence does not, however, deny the possibility that distinctive genetic elements occur in genetic substations; it only indicates that such genetic distributions, if observed, would have to be interpreted as consequences rather than as causes of the distinctive features of cellular regions.

Several studies have been directed to the question of cortical DNA. These are not examined in detail here. The first reports seemed to support the concept of cytoplasmic genetic depots, but more recent studies have been less positive. Several confounding problems exist. First, tritiated thymidine, which in most eukaryotes may be relied upon to go exclusively into DNA, seems to be sidetracked into RNA in some ciliates. The presence of label from this source requires appropriate controls to distinguish DNA from RNA. Second, the techniques for flaying ciliates are not perfectly developed, so that autoradiographed "skins" may contain various cytoplasmic or nuclear contaminants. Lysed macronuclei may accidentally contribute DNA to stripped cortex. Mitochondria, moreover, which certainly contain DNA, are regular, periodically distributed cortical components in ciliates and are difficult to dissociate from the cortex. If RNAase insensitive label appears in clean cortical preparations in a regular pattern similar to that of basal bodies, ultrastructural analyses are still required to see whether the label is associated with mitochondria rather than basal bodies. Thus far, the most fastidious studies fail to confirm the presence of DNA in the cortex, except that DNA associated with mitochondria.

Evidence suggestive of RNA associated with the basal bodies has, however, been obtained. Because of the multiplicity of RNA roles, its presence in basal bodies does not explain its function. Sequestered genetic messages may yet prove to be an aid in achieving regional specialization in an acellular organism. Even if this is so, however, the experimental evidence previously reviewed is as cogent with respect to RNA reservoirs as to DNA reservoirs. Their localization would first have to come about as a consequence of a prior localization of structure and/or function; their activities might, however, promote and help perpetuate those local specializations.

SUMMARY

The structural elements of the ciliate cortex provide the scaffolding for the insertion of new organelles. Because of this morphogenetic role of preexisting structure, the cell has some of the properties of an organismic crystal. Cells may reproduce in several different stable configurations, differing not at all in their molecular composition, but only in their pattern of organization.

All the architectural inertia of the cortex does not, however, depend on the visible structures, because patterns may persist under circumstances in which all the major cortical components have been dismantled. A more subtle molecular substrate for the pattern must persist in the absence of its manifestation.

This molecular substrate does not, however, rely on the template capacity of nucleic acids. If distinctive distributions of nucleic acid are eventually associated with the cryptic patterns, they will have to be explained as consequences rather than as causes of that pattern.

RECOMMENDED READING

Beisson, J. and T. M. Sonneborn. 1965. Cytoplasmic inheritance of the organization of the cell cortex in *Paramecium aurelia. Proc. Natl. Acad. Sci. U.S.*, **53**, 275–282.

Dippell, R. V. 1968. The development of basal bodies in Paramecium. *Proc. Natl. Acad. Sci. U.S.*, **61**, 461–468.

Frankel, J. 1973. Dimensions of control of cortical patterns in Euplotes:

the role of preexisting structure, the clonal life cycle and the genotype. *J. Exp. Zool.*, **183,** 71–94.

Grimes, G. W. 1973. An analysis of the determinative difference between singlets and doublets of *Oxytricha fallax*. *Genet. Res. Camb.*, **21,** 57–66.

Grimes, G. W. 1976. Laser microbeam induction of incomplete doublets of *Oxytricha fallax*. *Genet. Res. Camb.*, **27,** 213–226.

Hanson, E. D. 1967. Protozoan development. *Chem. Zool.*, **1,** 395–539.

Heckmann, L. and J. Frankel. 1968. Genic control of cortical pattern in Euplotes. *J. Exp. Zool.*, **168,** 11–38.

Nanney, D. L. 1966. Corticotype transmission in Tetrahymena. *Genetics,* **54,** 955–968.

Ng, S. F. and J. Frankel. 1977. 180° rotation of ciliary rows and its morphogenetic implications in *Tetrahymena pyriformis. Proc. Natl. Acad. Sci. U.S.,* **74,** 1115–1119.

Sonneborn, T. M. 1963. Does preformed cell structure play an essential role in cell heredity? In *The Nature of Biological Diversity.* (J. M. Allen, Ed.), McGraw-Hill, pp. 165–221.

Sonneborn, T. M. 1977. Local differentiations of the cell surface of ciliates: their determination, effects and genetics. *Cell Surface Rev.,* **4,** 829–856.

11

Cytonumerology,
Cytogeometry,
and
Positional Information

THE PHENOMENA DISCUSSED IN THE PREVIOUS CHAP-
ter demonstrate the regional specialization of ciliate protoplasm
and indicate that specialized structures and functions arise in
particular regions because of the positions of those regions in a
larger spatial organization. The means whereby cells enumerate,
deploy, and integrate their component parts are unknown, but the
presence of a dynamic regulatory system monitoring and
maintaining structural relationships is beyond dispute. Here we do
no more than offer illustrative examples of structural regulation in
a few ciliates.

A. The Numbers of Cortical Units in Tetrahymena

Tetrahymena is a relatively simple ciliate. It usually has 18–20
straight ciliary rows extending without interruption from the
anterior to the posterior end of its flame-shaped body (Fig. 11-1).
The subterminal oral apparatus provides a landmark identifying
the ventral surface and also prevents one, two, or occasionally
three of the ciliary rows (the *postoral rows*) from reaching the
apex. The ciliary rows are easily visualized by the use of silver
stains, either with the Chatton-Lwoff technique or the protargol
(silver proteinate) procedure. The protargol procedure in particular stains the basal bodies and permits them to be counted.
Simply counting the basal bodies—as indicators of cortical units—
can be informative with respect to the regulation of cortical
organelles.

The first question that arises is whether ciliary units are added
continuously during the cell cycle or only at a specific time. This
question is answered by counting the basal bodies in staged cells;
the stages of stomatogenesis provide a convenient reference for
much of the cycle (Fig. 10-3). The answer is that basal bodies are
generated throughout the cell cycle, although not at a perfectly

177

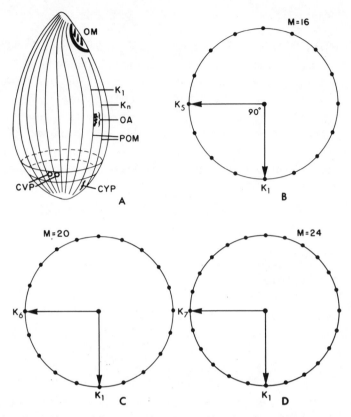

Fig. 11-1. A cytogeometric interpretation of the placement of the contractile vacuole pores (CVPs) relative to the first ciliary row (kinety 1, $K1$). (A) Right lateral view of surface features. $K1$ and Kn are the left and right postoral ciliary rows or meridians (POM). The cytoproct (CYP) is located just to the left of row 1. The contractile vacuole pores are located near the posterior end about one quarter of the distance around the cell to the right. The mean position of the CVPs may be identified by a constant central angle of 90°, using ciliary row 1 as the other reference point. The particular ciliary row representing the mean CVP position varies, depending on the total number of rows. (B) In cells with 16 meridians the mean position of the CVPs is at row 5. (C) Cells with 20 rows have the CVPs centered at row 6. (D) cells with 24 rows have a mean CVP position near row 7. Reprinted, by permission, from Nanney, D. L. 1966. Cortical integration in Tetrahymena. An exercise in cytogeometry. *J. Exp. Zool.*, **161,** 307–317. Fig. 1, p. 312.

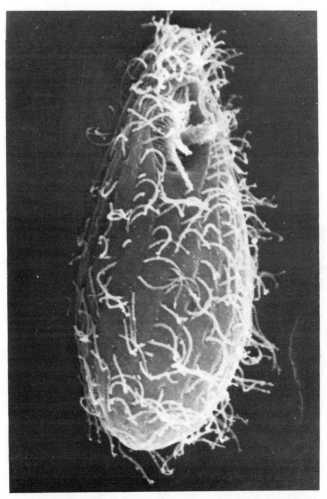

Plate VI. Scanning electronmicrograph of *Tetrahymena thermophila.* Ventral view showing the ciliary rows and the oral membranelles. Nelsen, E. M. and L. E. Debault 1978. Transformation in *Tetrahymena pyriformis*: description of an inducible phenotype. Magnification: 3000×. *J. Protozool.,* **25,** 113–119.

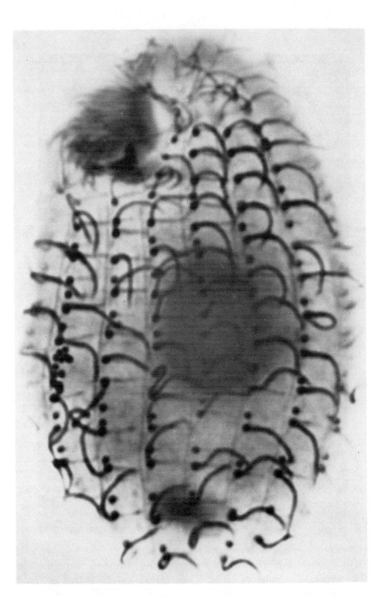

Plate VII. Protargol stained *Tetrahymena thermophila.* Ventral view showing the ciliary rows and the bases of the cilia. Magnification: 2300×. Courtesy of E. M. Nelsen.

constant rate overall. This behavior is different from that in Euplotes and some other ciliates in which all replication of basal bodies occurs just prior to cell division.

Another question is that of the anterior-posterior distribution of new basal bodies. This question can be answered because new basal bodies lack a ciliary shaft for a considerable period of time. Are new basal bodies equally likely to arise in association with any of the basal bodies in a row? The answer is negative. As in Paramecium new basal bodies tend to appear predominantly in the midregions of the cell.

The final question is whether any circumcellular pattern of basal body development can be discerned. The clearest evidence on this point comes from Kaczanowski's observations on severely starved cells. Under starvation conditions the cell may reduce its volume drastically and resorb many of its organelles. When feeding is again permitted, basal body formation occurs at a rapid rate, so that some old basal bodies are associated with several new basal bodies. These new basal bodies tend to be located in the midregion of the cell, as is noted earlier for normal growing cells; the new basal bodies are also concentrated on the ventral surface. Since each ciliary row must double its basal bodies in each cell cycle, the rate of production of new basal bodies on a cell cycle basis must be the same for all rows with equivalent numbers of units. However, during refeeding, and perhaps during stomatogenesis, a dorsoventral gradient of proliferation is apparent. Kaczanowski proposes that this coincidence of a temporal and two spatial proliferative gradients is responsible for the excessive production of basal bodies in the midventral region that initiates stomatogenesis.

The addition of basal bodies is thus seen as a process in which the rates of production of presumably identical organelles in different parts of the cell at different times in the cell cycle are carefully regulated. A different and puzzling kind of regulation is found when we consider the relationship between the numbers of basal bodies (at some prescribed time in the cell cycle) and the numbers of ciliary rows. Most Tetrahymenas have 18–20 ciliary rows, but exceptional individuals may be obtained with 16 or fewer rows or up to 30 or more. Is the number of basal bodies per row the same, regardless of the number of rows? If regulation occurred at the level of the ciliary row, the length of the cell (in numbers of ciliary units) would remain constant, and independent of row numbers, but the cells would become wider (in numbers of ciliary units) as the row number increased, changing the shape of the cell. The shape of the cell (width/length) could be preserved if as the number of rows increases, the number of ciliary units per row would also increase an appropriate amount. Neither of these relationships is found, however (Table 11-1). As the numbers of ciliary rows *increase,* the numbers of basal bodies per row *decline*; this compensatory regulation results in a constant number of basal

TABLE 11-1

Counts of basal body numbers in particular ciliary rows and estimates of total basal bodies in cells of *Tetrahymena thermophila*

The mean number of basal bodies per row decreases as the number of ciliary rows increases. The total number of basal bodies per cell shows no systematic change. Correlation coefficient (no. ciliary rows/ total basal bodies) = − .020 (nonsignificant).

Total Ciliary Rows/Cells	Number of Cells Counted	Mean Basal Bodies in Representative Rows			Estimated total Basal Bodies
		2	n	$n-2$	
16	1	52	30	49	764
18	16	53	26	48	852
19	8	54	25	46	865
20	6	39	20	36	706
21	6	43	21	40	819
22	3	45	18	39	862
23	3	39	20	36	797
24	1	45	23	37	959

Source: Nanney, D. L. *J. Exp. Zool.*, **178**, 177–182 (1971).

bodies per cell, regardless of the number of rows and regardless of the severe changes in the ratio of the width to the length of the organism.

This constancy of basal bodies should not be interpreted too strictly. We have noted cell cycle and nutritional effects on numbers of basal bodies. But even taking these into account, cells with the same numbers of ciliary rows may still show considerable variation around the mean. The coefficient of variation, $s/\bar{x}$, is about 0.10. This variability may be related to the variability in the amount of DNA, which is only approximately divided at cell division, and which is regulated only within certain limits (Chapter 8).

One other aspect of the quantitative regulation of basal body numbers merits brief consideration. Doublet tetrahymenas may be constructed with twice the normal number of ciliary rows by treating conjugating pairs with immobilizing antiserum. These doublet cells initially have 36–40 ciliary rows, but are unstable in this respect and lose ciliary rows quickly until 30–32 rows remain. This condition is reasonably stable and may persist for many cell generations, particularly if the doublet is perfectly symmetrical. Gradually, additional rows are lost, but the cell usually remains doublet in terms of its specialized organelles (oral apparatus, cytoproct, contractile vacuole pores) until it is reduced to about 25 ciliary rows. At this time, one of the sets of duplicate specialized organelles is repressed, though not simultaneously for the various organelles, and the doublet is transformed into a singlet.

An analysis of the basal bodies in doublets shows that doublets initially contain twice as many basal bodies as singlets and that the number of basal bodies per row is about twice as great as would be found in a singlet with the same total number of rows. Each half of the cell behaves as a unit and has a quota equal to the inventory of a singlet. As the doublet begins to transform, through the loss of ciliary rows and major organellar elements, the double-quota system also disintegrates, and a single reintegrated inventory is established.

One final kind of regulation of ciliary organelles also may be mentioned. In a recent survey of basal body numbers in some 17 species of the *T. pyriformis* complex, the total variation in the means of stage six half cells (just before cell separation) was from 246 to 486. Because of this limited difference between species and the variability even within a strain, basal body numbers are nearly useless taxonomically. The variation within a single cell cycle in a strain is about as great as that of the means of all the species in the entire species complex. Because the cilia, and the ciliary units, in this group of organisms are built on the same scale (indeed cilia are nearly uniform in cross section throughout the eukaryotes), the numbers of ciliary units are strongly indicative of organismic scale. The ciliated protozoa as a whole have a great diversity of scale, but the tetrahymenas show very little dispersion in spite of their large molecular distances (Chapter 5). They have had ample time and opportunity to evolve size variations. Their failure to diversify signifies stringent selective pressure to maintain a precise scale peculiarly suitable for this particular organic design.

B. The Position of the Contractile Vacuole Pores in Tetrahymena

The Tetrahymena organismic design is conserved not only in scale, but also in the positional relationships of its parts. Particular attention has been focused on the relative positions of the contractile vacuole pores (CVPs) and the stomatogenic ciliary row. Most cells have two CVPs located to the left of adjacent ciliary rows on the right side of the cell. One simple question concerning the CVPs is whether they are associated with particular rows or are located at some relative position around the circumference of the cell. The question may be answered again by recourse to populations of cells with different numbers of ciliary rows. If the CVPs are associated specifically with rows 5 and 6, for example, they will remain associated with rows 5 and 6 whether the cell has 16, 18, 21, or 24 total rows. If, on the other hand, the CVPs are located at a constant proportional distance from the stomatogenic row, their row positions will have to shift as the total number of rows increases. This adjustment of row position in order to maintain a constant cellular position is in fact observed (Fig. 11-2) and occurs in all the species examined. As in the case of stomatogenesis, no fixed relationship exists between a particular ciliary row and a major structure or activity. The ciliary rows are equipotential, and their functions are determined by their spatial relationships.

Doublet cells provide an interesting probe of the mechanism of CVP positioning. In doublets, two sets of CVPs are formed, on opposite sides of the cell. Are they formed at the usual position, about a fourth of the distance around the cell? The answer is negative. The CVPs are located only one quarter of the distance from one stomatogenic row to the next (Fig. 11-3). When a doublet transforms with the loss of one of the stomatogenic rows, the CVPs remaining move out twice as far from the remaining stomatogenic row as before and the duplicate set of CVPs is lost.

Just as the scale of the tetrahymenids is remarkably conserved, so that very little size variation occurs among the species, so also is the basic cytogeometry maintained in the related species. In a survey of 19 species of the *T. pyriformis* complex, the mean position of the CVPs varied between about 20 and 30% of the cell's circumference to the right of the stomatogenic row (Fig. 11-4). Most of the species had their CVPs between 21 and 25% of the circumference to the right of the stomatogenic row. No explanation has been offered for this location of the CVPs on the side of the cell, much less on the *right* side of the cell. The asymmetry almost certainly has functional significance, however, and it may be related to the other organellar asymmetries. The ciliary unit is, of course, asymmetric and the ciliary rows composed of files of ciliary units are asymmetric. The oral membranelles are asymmetric, with the single undulating membrane opposed to the three membranelles of the AZM. The gullet opening at the posterior end

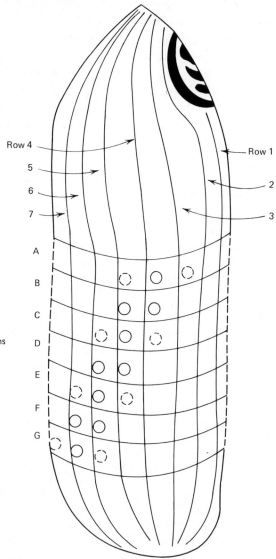

Row 4
5
6
7

Row 1
2
3

A
B
C
Patterns
D
E
F
G

186

of the oral apparatus is also asymmetric, directed toward the cell's left margin. Food vacuoles therefore begin their cytoplasmic circuit in a prescribed asymmetric pattern. Perhaps we should not be surprised that the water balance organs are asymmetrically placed. We should remember, however, that the "regulation" of CVP position we are discussing here is regulation by selective forces over evolutionary time, rather than regulation by developmental maneuvers. The evolutionary judgment on cellular design must, however, be achieved through epigenetic mechanisms. We are as incapable of explaining the means as we are the ends.

C. The Position of the Contractile Vacuole Pores in Chilodonella

Ciliates of the genus Chilodonella are very different from the modified cylinders of Paramecium and Tetrahymena (Fig. 11-5). The animals are flattened dorsoventrally, and the cilia are

←——————————————————————————————

Fig. 11-2. Positions of contractile vacuole pores on cells of *Tetrahymena thermophila* with different numbers of ciliary rows. Pattern A includes cells with one CVP to the (cell's) left of row 4, or three CVPs associated with rows 3, 4, and 5; the average distance from row 1 is 2.75 intermeridianal spaces. Pattern B consists of cells with CVPs to the left of rows 4 and 5, at an average distance of 3.25 spaces.

Total Ciliary Rows per Cell	Percentage of Cells With Pattern							Mean Distance From Row 1	Fraction of Circumference
	A	B	C	D	E	F	G		
17	2.2	86.7	8.8	2.2				3.31	0.195
18	0.4	41.9	17.0	40.7				3.74	0.208
19		2.2	5.2	92.1	0.5			4.20	0.221
20		0.8	0.8	95.9	1.6	0.8		4.25	0.213
21			0.7	93.8	3.4	2.1		4.28	0.204
22				71.2	16.9	11.9		4.45	0.202
23				17.6	29.4	47.1	5.9	4.96	0.216

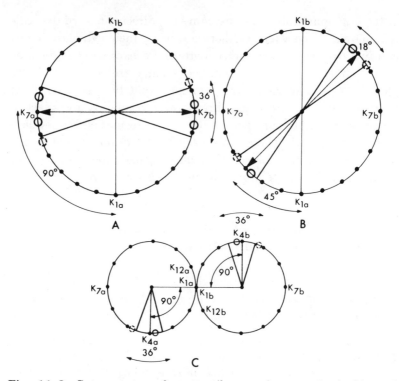

Fig. 11-3. Cytogeometry of contractile vacuole pores in doublets of *Tetrahymena thermophila.* (A) Predicted positions (and numbers) of CVPs in doublets if the cell measures the entire circumference of the cell as a reference. (B) Observed positions (and numbers) of CVPs in doublets, indicating that the doublet consists of two nonoverlapping morphogenetic fields. (C) An alternative geometric construct equivalent to B. Reprinted, by permission, from Nanney, D. L. 1966. Cortical integration in Tetrahymena. An exercise in cytogeometry. *J. Exp. Zool.,* **161,** 307–317. Fig. 1, p. 312.

restricted to ciliary rows on the ventral surface. The oral apparatus is ventrally located about a third of the distance from the anterior end. The multiple contractile vacuole pores are also on the ventral surface, scattered about in an apparently haphazard way and differing somewhat from individual to individual.

This apparently random distribution is rationalized, however, by Kaczanowska in terms of a geometric morphogenetic model.

The original coherent morphogenetic field of the ventral surface is dissolved prior to fission by the development of a second morphogenetic center midway between the old oral apparatus and the caudal margin of the cell. The old center (the oral apparatus for the anterior cell) and the new center (corresponding to the presumptive position of the oral apparatus for the posterior cell) serve as points of reference for the placement of CVPs. The new CVPs arise preferentially at prescribed radial distances from the centers, these distances being designated as radius 1, radius 2, . . . radius 5 and corresponding to actual distances (in strain X) of approximately, 14, 18, 22, 26 (or 27), and 30 micrometers.

These observations suggest a system of standing morphogenetic waves of constant amplitude associated with two centers. The

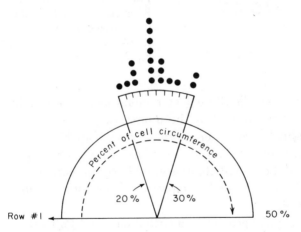

Fig. 11-4. The distribution of mean positions of contractile vacuole pores in 19 species of the *Tetrahymena pyriformis* complex. All the positions are within 20-30% of the distance around the cell, and most are within 23-25%. Based on data from Nanney, D. L., D. Nyberg, S. S. Chen, and E. B. Meyer. 1980. Cytogeometric constraints in Tetrahymena evolution: contractile vacuole pore positions in 19 species of the *T. pyriformis* complex. *Am. Nat. (in press)*.

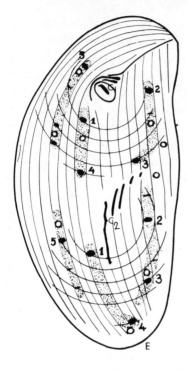

Fig. 11-5. The ventral surface of *Chilodonella cucullulus,* stock X. The curved lines represent the ciliary rows. The circular structure at C_1 is the old oral apparatus; C_2 represents the site of the new oral apparatus and a second morphogenetic center. The small open circles are the old contractile vacuole pores, destined to be replaced by new CVPs, identified as black circles. These new CVPs arise preferentially at multiples of a radial distance from C_1 or C_2 and at an intersection with one of the ciliary rows. Reprinted, by permission, from Kaczanowska, J. 1974. The pattern of morphogenetic control in *Chilodonella cucullulus J. Exp. Zool.,* **187,** 47–62. Fig. 5, p. 52.

CVPs are not distributed at random along these radii, however, but are concentrated at the intersections between the radii and longitudinal streaks of morphogenetic capability, which follow generally certain ciliary rows. These intersections provide more potential positions than are realized in actual CVPs. Some constraints seem to be imposed on the total number of CVPs developed.

The regular geometric distribution of CVPs in the ventral topography becomes distorted and eventually lost as fission is completed and the cells begin growing. The cytogeometric devices are more concerned with the initiation of structures than with their mature distribution. The Chilodonella analysis shows that the placement of organelles on a flat surface may be achieved by mechanisms similar to those responsible for the placement of structures on the surface of a cylinder.

D. The Position of the Oral Apparatus in Tetrahymena

A final example of positioning of structures is again taken from studies on Tetrahymena and is concerned with the placement of the new oral apparatus prior to fission. Under ordinary circumstances the oral primordium in Tetrahymena arises at a position that may be described in either of two ways. One may say that it arises at some fraction of the distance between the old oral apparatus and the posterior margin of the cell. Or, one may say that it arises at a a fixed distance from the old oral apparatus. Either description is usually accurate, but the inferences concerning the means whereby the cell places the primordium are different.

Several methods have been used in an attempt to answer this question of how the cells place the primordium. Lynn and Tucker have employed the somewhat unusual division pattern in *Tetrahymena corlissi* to explore the question. When starved cultures of this species are given food the cells do not divide for 12–15 hours; then a somewhat synchronized burst of cell division occurs. This division is followed very quickly, usually within 1 hour, by a second division of the same cells. The "first divider" is slightly larger (72.5 micrometers in length) than a log divider (69.4 micrometers) and much larger than a "second divider" (52.6 micrometers). Thus, unlike some Tetrahymenas, cells of the same strain but of very different sizes may be examined at the time of division. One may now ask whether in these diversified cells the new mouthparts develop at a constant distance from the old oral apparatus, or at some proportional distance from it. The answer (Fig. 11–6) is that the distance between the mouthparts is a varia-

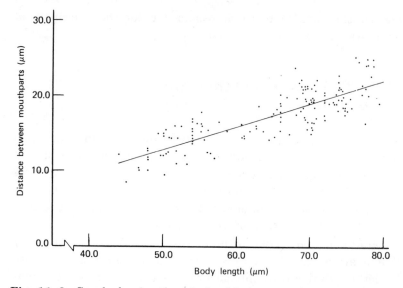

Fig. 11-6. Graph showing the relationship between the mouthparts (*d*) and body length (*l*) for 150 silver stained dividing cells of *Tetrahymena corlissi* (50 each of first and second poststarvation dividers and 50 log dividers). The line fitted by linear regression analysis has the equation *d* = 0.3055 (*l*) − 2.474, where *d* and *l* are in micrometers. Reprinted, by permission, from Lynn, D. H. and J. B. Tucker 1976. Cell size and proportional distance assessment during determination of organelle position in the cortex of the ciliate Tetrahymena. *J. Cell Sci.,* **21,** 35–46. Fig. 2. p. 39.

ble and that it is linearly related to the total cell length. The cell does not mark a constant distance from some cytological landmark in its placement of a new cellular structure, but measures some relative distance between two cellular locations. The analysis does not identify with certainty the cytological elements used as references, that is the anterior end of the cell versus the midpoint or the posterior margin of the old apparatus, or the posterior end of the cell versus the posterior end of the cytoproct. The analysis does indicate that the morphogenetic system is concerned with proportional rather than absolute intervals between organelles.

This analysis is challenged by observations on *T. thermophila*, in which cellular parameters are changed by genic substitution rather than nutritional manipulation. A morphological mutant studied by Doerder and coworkers and designated as *conical* has a drastically changed shape (Fig. 11-7); it is about the same size as the wild type cell, but is shorter, broader, and less symmetrical in the anterior-posterior axis. Doerder and coworkers asked whether the oral primordium in these cells occurs at a proportional or an absolute distance from the old oral apparatus and concluded that the absolute distance was identical in both cell forms. The consequence of this placement is that the new oral apparatus and the fission furrow are located nearer the posterior end of the mutant cell than the anterior end. The anterior daughter is systematically larger than the posterior daughter. This inequality in size must be compensated by different growth intervals before the next cell division, when inequality is again imposed.

In an attempt to reconcile these conflicting interpretations as to the manner in which cells place structures during morphogenesis, Lynn undertook a more extensive analysis of the *T. thermophila* material. First he demonstrated that, although dividing cells of this species are not as variable in size as those in *T. corlissi*, they nevertheless show considerable variation. Moreover, the positions of the new oral apparatus—measured as the distance between the old and new mouthparts—are not constant even within a strain. The larger cells have new primordia at greater distances than do smaller cells. When data from wild type and conical cells are plotted together (Fig. 11-8), the relationships between length and primordium distance in the two strains are clearly different, even though the mean absolute distance is about the same.

Lynn then showed that these data could be superimposed (Fig. 11-9) by plotting the distance between the mouthparts against not the cell length, but the product of the cell width times the cell length. This latter figure is a rough measure of surface area. The coincidence of the curves when they are plotted against a surface measure rather than a linear measure suggests that the placement

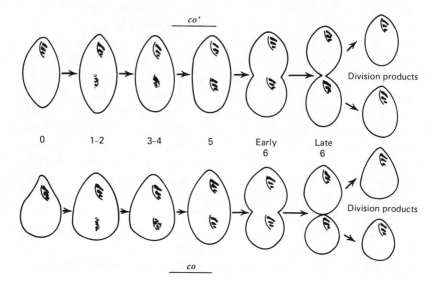

Fig. 11-7. Stomatogenesis in wild type (co^+) and a conical (co) mutant of *Tetrahymena thermophila*. The nondividing cells (stage 0) can be distinguished by their shape; the conical cells are shorter and broader (see below). At division (late stage 6) the two daughters of wild type cells are the same size and shape, but the two daughters of a conical cell are very different; the anterior daughter is the same length as the normal daughters, but the posterior daughter is shorter and more rounded. Apparently, the new oral apparatus in conical cells is located at the normal distance from the old apparatus, but at a much shorter distance from the posterior end. Figure kindly supplied by J. Frankel.

Stage	Genotype	Length L	Width W	W/L
0	co^+/co^+	36.6	20.2	0.56
	co/co	32.9	21.9	0.68
Late 6	co^+/co^+ anterior	26.2	17.8	0.68
	co^+/co^+ posterior	27.3	17.7	0.65
	co/co anterior	26.1	20.1	0.78
	co/co posterior	20.4	18.3	0.90

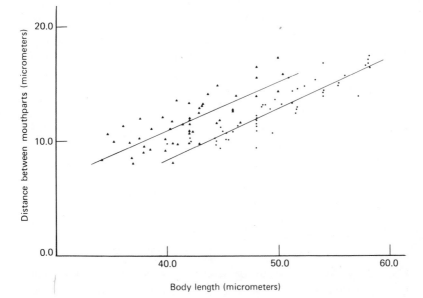

Fig. 11-8. The relationship between the distance between the mouthparts (d) and the body length (l) for 50 dividing organisms of co^+ and 50 co dividers. The lines fitted by linear regression analysis have the equations, $d = 0.477$ $(l) - 6.96$ for co^+ and $d = 0.447$ $(l) - 9.38$ for co. (●) co^+; (▲) co. Reprinted, by permission, from Lynn, D. H. 1977. Proportional control of organelle position by a mechanism which similarly monitors cell size of wild type and conical form-mutant Tetrahymena. *J. Embryol. Exp. Morphol.*, **42**, 261–274. Fig. 2, p. 266.

of the oral primordium is not a simple matter of assessing a proportion of a distance between two points, but involves a more complex evaluation of the cortical *area*. The abnormal shape of the conical cell has led an ordinarily adequate mechanism into inappropriate morphogenetic maneuvers.

Lynn further shows, in support of the significance of this global assessment, that even when the *T. corlissi* data are co-plotted with the *T. thermophila* data, a reasonable congruence is observed. The

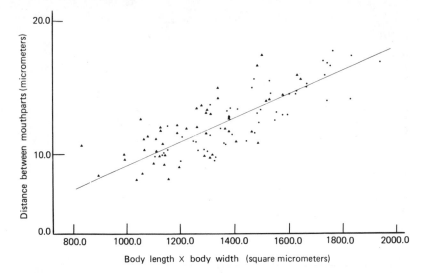

Fig. 11-9. The relationship between the distance between the mouthparts (d) and the body length–body width product ($l \times w$), an estimate of cell size or surface area for 100 dividing organisms (50 each of co^+ and co). The best line fitted by linear regression analysis has the equation $d = 0.00907 \, (l) \, (w)$. (●) co^+; (▲) co. Reprinted, by permission, from Lynn, D. H. 1977. *J. Embryol. Exp. Morphol.*, **42**, 261–274. Fig. 3, p. 267.

relationship between the distance for the primordium and the cortical area seems to be conservative in the tetrahymenids.

SUMMARY

Although certain liberties are allowed, ciliate organismic design is severely constrained in certain respects, particularly in the spatial relationships of the specialized organelles and in the inventories of organellar elements.

Because of their sizes, the scale of the architectural program in ciliates is similar to that in metazoan embryos, and the phenomena of relative placement of parts are remarkably parallel. Since ciliates are not cellularized, but manifest the capacity for relative placement of parts in a positional field, cellularization seems not to be an essential element in the manifestation of positional information. In ciliates, as in higher forms, the positional constraints are chiefly operative at the primordium stage and are relatively unimportant after morphostasis is achieved.

RECOMMENDED READING

Doerder, F. P., J. Frankel, L. M. Jenkins, and L. E. DeBault. 1975. Form and pattern in ciliated protozoa: analysis of a genic mutant with altered cell shape in *Tetrahymena pyriformis,* syngen 1. *J. Exp. Zool.,* **192,** 237–258.

Frankel, J. 1974. Positional information in unicellular organisms. *J. Theor. Biol.,* **47,** 439–481.

Frankel, J. and L. M. Jenkins. 1979. A mutant of *Tetrahymena thermophila* with a partial mirror-image duplication of cell surface pattern. II. Nature of genic control. *J. Embryol. Exp. Morphol.,* **49,** 203–227.

Jerka-Dziadosz, M. 1974. Cortical development in Urostyla, The role of positional information and preformed structures in formation of cortical pattern. *Acta Protozool.,* **12,** 239–274.

Jerka-Dziadosz, M. 1977. Temporal coordination and spatial autonomy in regulation of ciliary pattern in double forms of a hypotrich, *Paraurostyla weissei. J. Exp. Zool.,* **200,** 23–32.

Kaczanowska, J. 1974. The pattern of morphogenetic control in *Chilodonella cucullulus. J. Exp. Zool.,* **187,** 47–62.

Lynn, D. H. 1977. Proportional control of organelle position by a mechanism which similarly monitors cell size of wild type and conical form-mutant Tetrahymena. *J. Embryol. Exp. Morphol.,* **42,** 261–274.

Nanney, D. L. 1972. Cytogeometric integration in the ciliate cortex. *Ann. N.Y. Acad. Sci.,* **193,** 14–28.

Tartar, V. 1967. Morphogenesis in Protozoa. *Res. Protozool.,* **2,** 1–116.

Uhlig, G. 1960. Entwicklungsphysiologische Untersuchungen zur Morphogenese von *Stentor coeruleus, Ehrenberg. Arch. Protistenkd.,* **105,** 1–109.

Mutual Exclusion
and
Functional Inertia

12

THE MULTICELLULAR STATE RELIES ON A CELLULAR division of labor. Cells of the same genetic constitution, metabolizing in similar environments, differing mainly in their developmental histories, often have different compositions and perform different organismic services. Substantial evidence supports the belief that the differences characterizing determined and differentiated cells at least in some degree reflect conditional genetic expression rather than constitutional genetic composition, and that the differences are often maintained by epigenetic memory rather than by conventional genetic transmission.

Conditional genetic expression and epigenetic memory are by no means multicellular monopolies. The mechanisms must have been invented by the eukaryotic microbes that preceded the metaorganisms; they were perhaps used in different survival strategies, but they provided the preadaptational base for essential features of cellular specialization. The new features required for multicellularity were those involved in integrating the functions of the diversified cells, rather than in achieving diversity. Quite aside from such *a priori* reasoning, we know that modern microbes do rely heavily on differential gene action to circumvent commonly encountered environmental variables. The *lac* operon in *E. coli* is one of the few genetic elements whose regulation is reasonably understood; and systems of such operons appropriately linked can, in principle, account for most features of cellular diversification even in higher organisms. Yet *E. coli* is a haploid organism, and may be more committed to mutational variety than to regulatory adjustment to meet environmental challenges. The real specialists in genetic regulation are more likely to be found among the diploids, and especially the outbreeding diploids (Chapter 6). In such organisms one may well expect regulatory systems of even greater sensitivity and precision, and of great relevance to multicellular organisms.

A. Regulation of Immobilization
Antigens in *Paramecium primaurelia*

The best studied example of molecular regulation in the ciliates concerns the "immobilization antigens." When paramecia are injected into a rabbit, the rabbit responds with antibodies capable of immobilizing cells of the kind injected. The cilia adhere to each other and are no longer able to move the cells; adjacent cells may clump together. At high concentrations of serum the cells are killed. The antibodies are species and strain specific, generally affecting the strain injected, and close relatives, but not affecting other strains, or effective only at much higher concentration. These differing responses make it possible to classify the strains of a species, grown under uniform conditions, into a number of *serotypes.*

The specification that the different strains being compared be grown under uniform conditions is important, for sublines of the same strain grown under different conditions of feeding or temperature may respond differently to antisera. Cells grown at low temperature, for example, 12 °C may not be immobilized by antiserum prepared against the same strain at 22°. If the 12 °C cells are injected into rabbits they evoke antibodies capable of immobilizing them, but not the same strain grown at 22 °C.

One must identify, therefore, different antigenic types within a strain, as well as antigenic types that distinguish strains. A rationalization of the serotypes begins with the description of the antigenic repertoire characteristic of each of several strains (Fig. 12-1). Strain 60, for example, commonly manifests an antigenic type S when it is grown at low temperatures; it often is serotype G at intermediate temperatures, and at high temperature it expresses type D. Note that some lines of strain 60 manifest type S at 20 °C, while others manifest type G. Similarly, at 25 °C some sublines express type G and others type D. The temperature at which growth occurs places some limits on the serotypes expressed, but the temperature may not completely determine the

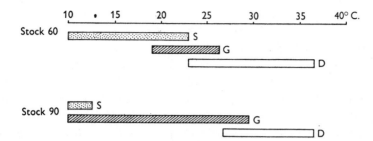

Fig. 12-1. Approximate temperature ranges for stable expression of three serotype loci in stock 60 and stock 90 of *Paramecium primaurelia.* Reprinted, by permission, from Beale, G. H. 1954. *Genetics of* Paramecium aurelia, Cambridge University Press. Fig. 8, p. 97.

serotype except under some limiting conditions. Thus sublines of stock 60 grown at 15 °C are always type S; those grown at 30 °C are always type D, but the serotypes of lines grown between 19 and 26 °C may not be predicted on this basis alone. Indeed sublines of type S and type G may be maintained almost indefinitely at 20° C and continue to breed true, and both G and D lines are stable at 25 °C. Raising or lowering the temperature, however, beyond certain limits may quickly lead to a change of serotype.

Each strain of *P. primaurelia* thus has an array of antigenic potentialities. Each cell characteristically expresses only one of the strain's potentialities; the chief exception occurs during the process of antigenic change. The different serotypes of a strain are *mutually exclusive* in expression. Changes in expression do occur, and these changes are usually triggered by environmental signals, so that certain environmental conditions are associated with certain serotypes. Type S is a low temperature serotype, for example. Changes of serotype within a strain are reversible changes, requiring often only a return to a previous environmental condition. The repertoire of antigenic potentialities is maintained intact throughout the changes from the expression of one type to another. All these changes may occur with little or no death, and with essentially 100% transformation of type.

Only after cataloguing the antigenic repertoire of a strain against common environmental variables is one ready to investigate strain differences. If we now consider the behavior of another strain, stock 90, under similar analysis, we find again a low temperature, a middle temperature, and a high temperature serotype. The serotypes in strains 60 and 90 differ, however, in two major respects. First, as shown, the conditions of stability of the types in the two strains are different. The low temperature serotype in stock 90 is observed only at 10–12 °C. Some 10 °C cultures, quite unlike those of stock 60, express the midrange serotype (G), which also has an increased range in the other direction.

The other difference between strain 60 and strain 90 serotypes concerns their immunological specificity. Antiserum against G cells in stock 60 is not very effective against G cells in stock 90, and vice versa. Similarly, stock 60D cells are readily distinguished from 90D cells. However, 60S cells cannot be distinguished from 90S cells, even though the S types of other strains are clearly different. The strain differences may involve both the kinds of antigens produced and the conditions for their production. We should, perhaps, indicate at this point that temperature is not the only environmental variable of consequence in serotype expression. Under certain "standard" conditions of culture, temperature may have a controlling function, but many other variables are important.

Once having defined the reaction pattern of each of the strains involved, we may begin an analysis of the genetic basis for strain differences. A line of strain 60 manifesting serotype G may be crossed with a line of strain 90 expressing G. Conjugation results in the establishment of identical genotypes in both conjugating cells (Chapter 7). In crosses between two homozygous strains the F_1 generation should be heterozygous for any alleles for which the parents differ. In the cross 60G $\times$ 90G, the F_1 individuals react equally to antisera against the two parents. These F_1 individuals may be further analyzed by allowing them to undergo autogamy.

At autogamy, each cell undergoes meiosis and self-fertilization and becomes homozygous for one allele of each pair of alleles it carries. When autogamy is induced in G hybrids between strains 60 and 90, about half of the cells give rise to clones expressing 60G, and half give rise to 90G clones. This is the result expected for the assortment of allelic genes. The difference in the specificity observed between 60G and 90G segregates as a simple Mendelian allelic difference.

In a similar manner one may examine the genetic basis for the D serotypes, or the S serotypes, by making crosses under appropriate environmental conditions between strains having distinctive serotypes. As in the previous case, one to one segregations at autogamy are obtained, showing again that these specific differences are due to allelic differences.

An important factor in evaluating these results comes from considering two or more kinds of serotypes simultaneously rather than separately. When a 60G/90G F_1 undergoes autogamy, yielding ½ (60G/60G) and ½ (90G/90G), how do these second generation lines behave with respect to D? The question is answered by raising the temperature so that D is expressed. When this is done, half of the 60G/60G F_2 lines are homozygous for 60D/60D and the other half are 90D/90D. If stocks 60 and 90 had distinguishable S antigens, which they do not, the F_2 could also be scored for this trait. Whenever two or more serotypes have been examined simultaneously, they have been found to show independent segregation in the F_2.

Earlier we note that strains differ in two major respects: in the specificities of the antigenic determinants and in the conditions for their expression. The analysis just discussed demonstrates that the specificities of the antigenic determinants are associated with single genetic loci, one locus for each serotype. The results do not yet inform us of the genetic control of serotype expression. Do the same loci controlling specificity also control expression? This question is answered by analyzing the patterns of expression of the F_2 clones produced. When an F_2 clone has the genotype 60D/60D,

60G/60G, its pattern of expression is similar to that observed in the original strain 60 parent. The same observation can be made for the F_2 genotype mimicking the other parent. Genes other than those associated with the antigenic structure itself may condition the expression of these antigens, but the influence of the other genes is apparently not strong in this case.

A more complex, but perhaps more interesting, situation arises when crosses are made between strains 60 and 90 while they are expressing different serotypes (Fig. 12–2). At 25°C all lines of stock 90 express type G, but some lines of stock 60 express D. Therefore we examine the F_1 resulting from the cross 60D × 90G. Here we must distinguish between the progeny of the two members of a pair; even though the two exconjugants have identical genotypes, they produce distinguishable progeny. The original stock 60 parent produces a type D clone, but one expressing both the 60D and the 90D antigen. The original 90 parent yields a type G clone with heterozygous expression. The new alleles entering a conjugant may come quickly to expression, but the decision as to which genetic locus is expressed is determined by the cytoplasmic environment. A conjugant producing the D antigen prior to conjugation continues to produce that type after conjugation, even though the genetic locus is changed from a homozygous to a heterozygous state.

These results establish the basic relations among the genes, the environment, the cytoplasm, and the history of the cell in determining the synthesis of these membrane proteins. Several different unlinked genetic loci provide an array of synthetic capabilities, and these loci are functionally associated in a system of mutual exclusion. Every cell must synthesize one, but only one, of the related molecules. The choice as to which molecule is produced is conditioned by the synthetic pattern already established. Under circumstances permitting the synthesis of any of two or more molecules, the molecule that has been produced is the molecule that continues to be produced. For this reason one may speak of *functional inertia* in describing the serotype system.

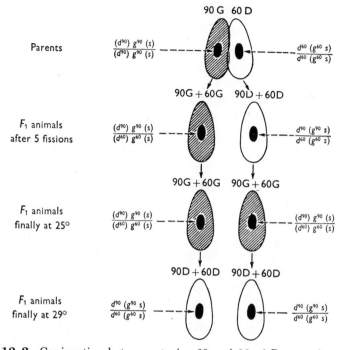

Fig. 12-2. Conjugation between strains 60 and 90 of *Paramecium primaurelia* while they are expressing respectively serotype D and serotype G. After a period of phenomic lag, the exconjugant from stock 60 continues to manifest serotype D, but with the allelic specificities of both stocks. Similarly, the exconjugant from stock 90 manifests both specificities of serotype G. Depending on the temperature, the D cells may transform to G or the G cells may transform to D, in either case they continue to manifest heterozygous expression. Reprinted, by permission, from Beale, G. H. 1954. *Genetics of* Paramecium aurelia, Cambridge University Press. Fig. 6, p. 95.

Functional inertia is responsible for a special kind of "epigenetic memory."

Generally, the system of functional inertia does not distinguish among the alleles at a locus. If at conjugation a new allele is introduced at a locus already expressed, the new allele is not discriminated against, but is equally expressed with the preexpressed allele. For this reason the regulatory system appears to be concerned with the regulation of expression of genetic loci rather than of the particular products of those loci. Some exceptions to this rule have been reported in *P. biaurelia* and in other species, and the basis for the irregular and reversible dominance in this species is still under investigation.

The functions of the immobilization antigens are unknown. The expression of some one of the products in all cells suggests that they have an essential role. The variability of the products in related species, and the wide polymorphism within species, suggest evolutionary lability, and hence some indifference concerning the details of molecular structure. Yet the cells clearly invest considerable effort in choosing for expression the one product that is in some respect the most appropriate.

The immobilizing antigens have been isolated and partially characterized. The antigens are very large protein molecules—on the order of 300,000 daltons, close to the largest polypeptides known. The genetic evidence provides no support for a multigenic control of their specificity and is consistent with exclusively protein, as opposed to glycoprotein, antigenic determinants.

B. Immobilization Antigens in *Paramecium tetraurelia*

All the species of the *P. aurelia* complex have systems of immobilization antigens somewhat similar to that described above, but only *P. primaurelia, P. biaurelia,* and *P. tetraurelia* have been subjected to intensive study. A brief discussion of the *P. tetraurelia* system is useful because this species manifests the phenomenon of functional inertia to a greater degree than does *P. primaurelia*. In the latter species, as is shown above, two or, rarely, three different antigenic types may be stably expressed by different representatives of a strain grown under the same conditions. In *P. tetraurelia* as many as six or eight stable types of a strain may be maintained simultaneously under a common growth routine.

The different antigenic types produced by a strain (A, B, C, . . .) are not equally stable under all growth conditions; indeed some, as in *P. primaurelia,* become unstable at high or at low temperatures. The stability is also affected by the kinds and amounts of nutrients provided. Nevertheless, the persistence of several distinctive types over a comparatively broad range of cultural conditions is notable.

The environment is critical for the manifestation of serotypes under certain circumstances, particularly when the serotype system has been "destabilized." Destabilization was initially discovered in analyzing the responses of cells to homologous antiserum. Cells immobilized by sublethal concentrations of antiserum may recover in an hour or two and begin swimming again, even in the presence of the antiserum. These "resistant" cells may be isolated, cloned, and studied further. Under some temperature and feeding conditions, and using a monotypic clone to begin with, for example, serotype G, all the cloned cells produced will be of the same serotype, for example, serotype D. The strain is transformed quantitatively from serotype G to serotype D by the application of anti-G serum.

This result may be modified by changing the temperature of treatment and/or growth, or the amount of food. Generally changes in environmental conditions change the course of transformation; some of the transformed clones may be type D, but others may be type A, B, and so on. The treatment with antiserum has altered the stability of the regulatory system, but the environment significantly affects the determination of the new stable types to be established.

Serotype destabilization may be achieved not only with antiserum treatment, but with a wide variety of physical and chemical treatments, such as radiation and metabolic inhibitors. In all these cases the course of transformation is strongly dependent on the environmental conditions.

The control of antigenic specificity has been analyzed, as in *P. primaurelia*, by genetic techniques. Each strain is first subjected to a thorough analysis of its serotype repertoire by destabilizing the cells and challenging them with diverse environments. Some strains have 14 or more different serotype capabilities, all associated in a system of mutual exclusion; one and only one is expressed by an individual cell at a particular time, and a type once expressed tends to continue its expression in the daughter cells.

Two strains may differ from each other in the specificity of similar serotypes; stock 51 has a serotype A similar to that of stock 47. We may refer to 51A and 47A. Two strains may, again as in *P. primaurelia*, differ somewhat in the optimum conditions of expression for the similar serotypes. Finally, some strains may produce serotypes not represented at all in other strains, at least under the environments examined. Any and all of these strain differences may be studied by breeding analysis. The conclusion from such studies is that the specificity of expression, and the capability of expressing a particular serotype at all, is associated with one particular genetic locus. The number of loci involved is at least as large as the number of serotypes known within a strain. These 14 or more genetic loci, though organized into a coordinated system of expression, are not clustered on the chromosomes. Indeed, no case of linkage among the serotype loci has been reported. Perhaps this result is not surprising in view of the large number of chromosome pairs in these species (40–50), but certainly it does not suggest a geographically localized polycistronic or multigenic complex.

The state of serotype expression, which is perpetuated through growth and successive cell divisions, is also maintained (again as in *P. primaurelia*) through the act of conjugation. A cross between a cell of strain 51 of serotype A and a cell of strain 47 of serotype D yields one exconjugant clone of type A and another of type D. Where suitable immunological tests are available the A clone can be shown to be expressing coordinately the allelic specificities of both parents; the D clone, similarly, expresses both D parental specificities. The A clone may be destabilized and transformed to a D type, with again a coordinant expression of both parental B specificities. The mutual exclusion system limits expression to one of the kinds of genetic loci in the system (A, B, C, . . .), but does not discriminate between the expression of allelic genes, even though their products are distinctive.

One final feature of the *P. tetraurelia* system should be noted. If cells of diverse serotype, for example A and D, are crossed under permissive conditions, one exconjugant usually expresses type A and one type D. If, however, a massive exchange of cytoplasm is induced during conjugation, the exconjugant clones are usually found to have identical serotypes, either both type A or both type D. The first result—persistence of phenotypic differences in "reciprocal crosses"—is a piece of classical evidence for "cytoplasmic heredity" (Chapter 15). The second result—disappearance of phenotypic differences when cytoplasmic mixture occurs—is considered confirmatory. Thus the inheritance of states of expression of the Paramecium serotypes has been considered as an example of cytoplasmic inheritance. The label, however, does not imply a mechanism. We return to this matter later. For the moment we consider that the antigen synthetic systems of the parental cells have "functional inertia," continuing in their previous patterns unless disrupted in some way. A mixture of two systems with different commitments and exclusion properties leads to conflict, disruption, and "destabilization." Which new stable synthetic system is established depends on the relative strengths of the opposing systems in the mixture, which may be conditioned by the environment. Even if the conflicting systems are equivalent in inertial force, the prevailing environment may condition the direction of exclusion. Certainly one may markedly affect the consequences of such crosses by manipulating the environment and hence directing the establishment of a stable state.

C. Immobilization Antigens in *Tetrahymena thermophila*

Tetrahymena strains also have the capability of provoking the formation of immobilizing antibodies when injected into rabbits. The homologies between the Tetrahymena and Paramecium systems are uncertain, particularly since the Tetrahymena antigens appear to be relatively small proteins—about 30,000 daltons in comparisons with 300,000 daltons for Paramecium.

Regardless of the question of homology, the analogies between the Tetrahymena and Paramecium systems are strong (Fig. 12–3). A strain of *T. thermophila* produces a characteristically different immobilization antigen at 15, 25, and 38°C. These antigens are referred to as L (low temperature), H (high temperature), and T (torrid). The precise temperature ranges for these serotypes, and the specificities of the serotypes, vary from strain to strain. Other serotypes may be evoked by particular treatments. Thus H cells

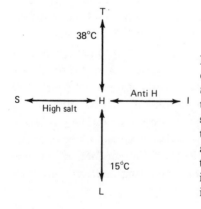

Fig. 12–3. Antigenic loci brought to expression in strains of *Tetrahymena thermophila* under different conditions. At 25°C in standard medium serotype H is expressed. Transformation occurs when cultures are placed at higher or lower temperatures, when the salt content of the medium is increased or when dilute anti-H serum is added.

treated with anti-H serum, even at concentrations that have no visible effect on motility, transform to type I (induced). Similarly, cells exposed to high salt concentration transform to a new antigenic type, S (salt), expressed only as long as the cells are maintained in this environment. The array of serotype states has never been exhaustively explored, but certainly it consists of more than the five now named.

The serotype system of Tetrahymena is thus like that in Paramecium, in that it is composed of a multiplicity of mutually exclusive states. And, as in Paramecium, in *T. thermophila* the expression of the serotype is strongly influenced by the environment, but not exclusively determined by it. Cells grown at 19 °C after having previously grown at 15 °C continue to manifest serotype L for long periods of time. However, cells grown at 19 °C after having previously grown at 25 °C continue to produce antigen H. Tetrahymena cells, like Paramecium cells, "remember" their previous environmental history and can be provoked to "forget" only by passing them into a less ambivalent environment.

The genetic basis for the multiplicity of serotypes in Tetrahymena is again like that in Paramecium. Allelic differences have been shown for serotypes H, T, and S, and these are unlinked. Recent studies by Doerder show that mutations affecting the expression of these loci may be obtained at still other loci. The regulatory system is not geographically localized in the genome.

Thus far convincing evidence of serotypic inertia through conjugation in Tetrahymena has not been provided. Crosses can be made between cells expressing different serotypes in permissive environments, and one report of cytoplasmic continuity of serotypes has been produced. In this case, however, the published evidence indicates that the conjugating pairs aborted without completing conjugation. More fully controlled crosses suggest that exconjugants from normal pairs have identical serotypes. The difference between this result with Tetrahymena and the usual result in Paramecium may reflect a difference in the amount of cytoplasmic mixing occurring at conjugation in these organisms. The transfer of pronuclei in both cases must involve some cytoplasmic transfer, and perhaps similar amounts in the case of organisms with similar pronuclear sizes. The *relative* amount of cytoplasmic transfer would likely be more significant in a small form such as Tetrahymena than in a large organism such as Paramecium. In any case McDonald has given evidence of considerable cytoplasmic exchange in Tetrahymena during conjugation.

The suggestion emerges, therefore, that conjugation in Tetrahymena is associated with cytoplasmic mixing and that this mixing, as in the case of induced mixing in Paramecium, leads to destabilization. The subsequent restabilization depends on the action of the preexisting serotype states, now mixed, and the critical environmental determinants, which are the same for both exconjugants. Both exconjugants then characteristically come to have the same serotype.

D. The Mechanism of Mutual Exclusion

The details of the mechanism for mutual exclusion and functional inertia of serotype expression in ciliates have not been discovered, but the probable class of mechanisms was perceived long ago by Kimball and by Delbruck. Delbruck's simple scheme (Fig. 12-4) provides two alternative biochemical pathways, each of which produces an intermediate product antagonistic to the other. If by chance or circumstance an intermediate, a2, reaches a concentration effectively higher than b2, a2 will inhibit the pathway producing b2 and thereby increase the production of a2. Two alternative conditions are possible, that producing product a3 and that producing b3. If the balance is tipped in favor of one pathway or the other, a metastable condition will be established and maintained against considerable environmental fluctuation with respect to factors

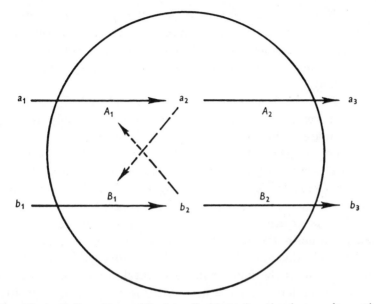

Fig. 12-4. Delbruck's model of a cell able to function in two alternative ways. Reprinted, by permission, from Beale, G. H. 1964. *Genetics of Paramecium aurelia*, Cambridge University Press. Fig. 9, p. 119.

affecting differentially the rates of reaction in the two sequences, or the supplies of precursors or reactants.

One of the better understood examples of functional inertia concerns the *lac* operon in *E. coli.* Novick and Weiner showed that wild type *E. coli* cells can be maintained in either of two stable states in the same environment, provided that the environmental supply of inducer is maintained at a certain level. More particularly, the *lac* operon in a strain of *E. coli* may be massively induced to synthesis by the artificial and gratuitous inducer TMG (thiomethyl β-galactose) in a concentration as low as $5 \times 10^{-4} M$. At a concentration of $5 \times 10^{-5} M$, essentially no induction occurs. Once a culture is induced, however, it will continue to produce β galactosidase in the presence of $5 \times 10^{-5} M$ TMG. Hence cells with a history of recent exposure to a threshold concentration of inducer will maintain their functional difference from cells not so induced. The memory of the historical event was followed in a chemostat for 200 cell generations and presumably could persist indefinitely.

The mechanism for maintenance of the induced state of the *lac* operon is well understood. The *lac* operon codes not only for the β-galactosidase, but also for a "permease," a membrane bound protein that promotes the accumulation of certain sugars (and inducers) against a gradient. A cell equipped with an ample supply of galactoside permease will concentrate TMG to a concentration inside the cell that is 100 times the concentration outside the cell. In the absence of the permease the internal and external concentrations of the inducer are equivalent.

For an induction of the *lac* operon to occur, the internal concentration of the inducer needs to reach a level of about $5 \times 10^{-4}M$. After induction has occurred, the internal concentration remains above $5 \times 10^{-4}M$, provided the external level does not fall too low. Even if the external level is allowed to drop to $5 \times 10^{-6}M$, below the level of initial induction, the internal concentration remains high enough to maintain the function of the operon and to insure the presence of permease concentrating the inducer.

The details of the *lac* metastable states are more complex than those proposed by Delbrück and are concerned with protein synthesis rather than with a biochemical pathway. The common feature is the ability of a product of a process to promote that process, a phenomenon designated as positive feedback. The product may be a metabolic intermediate, as in Delbrück's proposal, or a protein product of a polycistronic operon, but the systems acquire by virtue of positive feedback the capacity to stabilize in more than one functional state. Their stability is not in principle, or even in practice, irreversible, but it provides long term continuity of functional states.

The *lac* operon has two metastable states, and in the present consideration does not interact with other operons. Jacob and

Monod have shown, however, how interactions among operons can yield compound polyoperonic complexes with interesting features. Where such complexes provide positive feedback they can potentiate multiple metastable states with mutual exclusion.

Our understanding of the ciliate serotype systems is still primitive, but the systems must be complex enough to account for several geographically separated operons, mutual exclusion of expression, and metastability of state.

SUMMARY

Genetically identical ciliates not only manifest alternative architectural forms, but also different physiological states. They have a marked capacity to adjust to changing environmental circumstances by altering their synthetic activities. The different physiological states—patterns of molecular synthesis—like the patterns of structural organization, have inertial force. The physiological state may persist after the environmental stimulus is only a memory, but may nevertheless be altered again by an appropriate environmental provocation.

The most fully studied examples of demonstrably reversible physiological states are those involving the immobilization antigens. These systems consist of several widely scattered genetic loci associated in an organization of mutual repression; the synthesis of any one of the proteins excludes the synthesis of any other. The mechanism of perpetuation probably involves a feedback feature, but the level of the feedback loop and the means of mutal exclusion have not been discovered.

RECOMMENDED READING

Beale, G. H. 1952. Antigen variation in *Paramecium aurelia*, variety 1. *Genetics*, **37**, 62–74.

Beale, G. H. 1957. The antigen system of *Paramecium aurelia. Int. Rev. Cytol.,* **6,** 1–23.

Beisson, J. 1977. Non-nucleic acid inheritance and epigenetic phenomena. *Cell Biology: A Comprehensive Treatise* (L. Goldstein and D. M. Prescott, Eds.) Vol. 1, Academic, pp. 375–421.

Doerder, F. P. 1973. Regulatory serotype mutations in *Tetrahymena pyriformis. Genetics,* **74,** 81–106.

Finger, I. 1974. Surface antigens of *Paramecium aurelia.* In *Paramecium: A Current Survey* (W. J. van Wagtendonk, Ed.), Elsevier, pp. 131–164.

Gibson, I. 1970. Interacting genetic systems in Paramecium. *Adv. Morphol.,* **8,** 159–208.

Nanney, D. L. and J. M. Dubert. 1960. The genetics of the H serotype system in variety 1 of *Tetrahymena pyriformis. Genetics,* **45,** 1335–1349.

Sommerville, J. 1970. Serotype expression in *Paramecium. Adv. Microbiol. Physiol.,* **4,** 131–178.

Sonneborn, T. M. 1948. The determination of hereditary antigenic differences in genically identical Paramecium cells. *Proc. Natl. Acad. Sci., U.S.,* **34,** 413–418.

Sonneborn, T. M. 1950. The cytoplasm in heredity. *Heredity,* **4,** 11–36.

Macronuclear Differentiation and Random Karyonidal Determination

13

THE ROLE OF THE NUCLEUS IN CELLULAR DIFFER-
entiation is a persistent and still incompletely answered question.
Classical studies such as those of Spemann on newts indicated
that irreversible nuclear changes were not an essential feature of
development. Although most such studies were limited to amphibians, and to early stages prior to embryonic determination,
they were sometimes generalized to later stages and to less regulative organisms. Biologists were forced to consider the possibility of
nuclear constancy in development, not only with respect to the
genetic content of somatic cells, but also with respect to their
developmental potentialities and even their functional states. In
the absence of evidence *for* nuclear changes in development, the
dogma of nuclear equivalence became widely accepted. Finally,
however, the techniques of nuclear transplantation pioneered by
Briggs and King permitted the examination of nuclear capabilities
at later stages of development, after embryonic determination had
occurred. The results of these studies are generally agreed upon,
but their significance is still at issue. The developmental
capability of somatic nuclei, tested by injection into enucleated
eggs, declines progressively after the onset of gastrulation, at different rates in Rana and Xenopus, the two frog genera most extensively studied. Some nuclei transplanted from certain tissues of
Xenopus larvae can provide apparently normal development,
although the production of fertile adults from such nuclei is
extremely rare. Differentiated nuclei from adult tissue can sometimes give rise to an array of embryonic structures, indicating
pluripotency, but never to completely normal embryos. One may
explain away the failures of development with older nuclei on
technical grounds—the progressively smaller sizes of the cells, or
the lower probabilities of obtaining nuclei at the appropriate stage
of the cell cycle, for example. The fact remains that the observed
developmental capacity of the nuclei declines with developmental
stage, and the complete reversibility of the nuclear changes in
adult cells has not been convincingly demonstrated. We must

221

consider seriously the possibility that irreversible nuclear changes do occur in the somatic lineages of amphibians.

In recent years the question of the reversibility of developmental changes of mammals has also arisen, but in a somewhat different context. Particularly, somatic cells from cell cultures have been introduced into mouse blastocysts before implantation, and in some cases these somatic cells are incorporated into mosaic embryos and give rise to functional cells in a diversity of tissues. The tests do not distinguish the developmental plasticity of the somatic cell as a whole from that of its nucleus, but if the whole cell is pluripotent, then the nucleus must also be pluripotent. The question remains, however, as to whether the somatic cell is *totipotent,* that is, able by itself to generate an entire normal developmental program. Thus far no one has obtained total mammalian embryos derived from somatic cells. An alternative way to demonstrate totipotency would be to show that a somatic cell can be converted into a functional germ cell by sending it through a developmental program in a mosaic mouse. If a mosaic mouse produced normal eggs or sperms from a somatic cell ancestor, the reversibility of nuclear changes would be indicated. The evidence for such transformations is still equivocal; either the cell lineages used for the somatic parent are questionably "somatic" (as in teratocarcinoma lines) or the gametes are not functional. Again, the constancy of nuclear composition during development must be questioned, in mammals as well as in amphibians.

Finally, we should note that some kinds of genetic change certainly do occur during development. Earlier, we mention the phenomena of chromatin diminution and of genetic amplification (Chapter 8), which are programmed genetic changes associated with developmental processes both in ciliates and in some higher organisms. We should not neglect either the implications of the nuclear changes associated with the development of the immune response in mammals. Some of these "editorial" changes in the genetic transcripts may be reversible under some circumstances, but their general reversibility is doubtful. If genetic alterations are associated with differentiations in the immune system, they may also occur in other specialized tissues, though this has not yet been established.

The distinction between structural and functional nuclear states is not easily made in practice, however different they appear in concept. The genetic changes at the H locus ("phase variation") in Salmonella and at the mating type locus in yeasts provide analogies of possible application in development, because they violate our preconceptions about the nature of genetic modifications. We still think of precision of induction and reversibility of state as indications of functional alterations, but we can now easily imagine the programmed insertion or removal of a particular base or base sequence in a DNA segment at a specific time in development, which can render operative a previously silent operon. If a genetic alteration can be introduced specifically it might also be reversed specifically. Although structural and functional changes are not easily distinguished it is tempting to believe that irreversible changes (within the limits of observation) are more likely to involve structural changes than are regularly reversible changes. The ciliates provide notable examples of both kinds. Earlier we introduced the immobilization antigen systems (Chapter 12), whose states of nuclear expression may be environmentally manipulated at least within certain limits. Here we are concerned with other nuclear changes that have not thus far been found to be reversible, though environmental influences are important in their inception.

A. Macronuclear Determination

The first irreversible nuclear change in a ciliate is the induction of the macronuclear anlagen following fertilization. The zygote nucleus undergoes one or more mitotic divisions, and some (usually half) of these genetically identical nuclei become macronuclei, while the others remain as the germinal micronuclei. The nuclear fate is closely correlated with the nuclear position at the critical time of determination. In Paramecium the nuclei at the posterior end of the cell are induced to become macronuclei; in Tetrahymena those at the anterior end become macronuclei. The significance of the cytoplasmic location is supported by the observation that centrifugation of the conjugants at the time of nuclear induction can change the fate of the nuclei, presumably by dislocating the nuclei from their normal cytoplasmic environments. Exconjugants are produced with either too many or too few macronuclei. Apparently the cortical plasm in ciliates plays a role similar to that in insects; nuclei associated with some cortical regions retain their germinal character, but those associated with other regions begin an epigenetic program.

The irreversibility of macronuclear development is suggested by the observation that permanent amicronucleate strains develop in many ciliates. Some ciliates (such as *T. thermophila*) require micronuclei to continue in division, but in others (such as *P. tetraurelia*) the micronuclei are dispensable for a period of time at least. Once lost the micronuclei are not ordinarily regenerated, even though the macronucleus seems to try to compensate for the loss. In Tetrahymena, when conjugation involves an amicronucleate cell, the macronucleus extends a "finger" into the region of nuclear transfer and touches the membrane at the usual point of pronuclear transfer. In *P. bursaria* Schwartz has described the generation of an abnormal micronucleus in an amicronucleate clone presumably by a constriction from the macronucleus. Ammermann describes in detail the origin of "pseudomicronuclei" from macronuclei in cells of *Stylonychia mytilus* whose micronuclei were destroyed by radiation. In none of these cases, however, is the macronucleus able to compensate fully for the lost micronucleus. The pseudomicronuclei cannot participate normally in conjugation. The pseudomicronuclei of Stylonychia indeed contain many tiny chromatin elements, as would be expected in a form whose genomic tapes are completely edited and to a large degree erased during macronuclear development (Chapter 8). Even in the case of *T. thermophila*, thought to have less extensive nuclear alterations during macronuclear development, some irreversible structural changes must occur.

B. Mating Type Determination in *Paramecium primaurelia*

The regulation of mating, and hence the regulation of mating *types*, is an essential feature of ciliates' ecogenetic strategy (Chapter 5). The number of mating types in a species influences the probability of making the most, genetically, of a chance encounter with a stranger. The length of time mating is deferred after conjugation affects the genetic distance between the cells that mate. These and other aspects of mating require regulation of the synthetic activities of the macronucleus. We are here concerned, however, with a special kind of nuclear regulation that permits diverse mating types to occur within a single synclone and hence permits mating among genetically equivalent sublines. This inbreeding adaptation was first described by Sonneborn in *P. primaurelia*.

Sonneborn observed that isolated lines of *P. primaurelia* usually breed true during vegetative growth and that some are of the Odd mating type (I) and others are of the Even mating type (II). When conjugating pairs are isolated and allowed to give rise to *synclones*, most such cultures—derived from single conjugating pairs and genetically alike—contain some individuals of Even mating type and some of Odd. Because individuals isolated from young synclones yield pure Even or pure Odd clones, the diversities must arise early in the development of the synclone.

The establishment of hereditarily diverse lineages among the population of genetically identical cells was investigated by means of a *vegetative pedigree analysis* (Fig. 13-1). When the conjugating cells separate, each is transferred by a micropipette into a separate drop of culture medium. When the cells undergo their

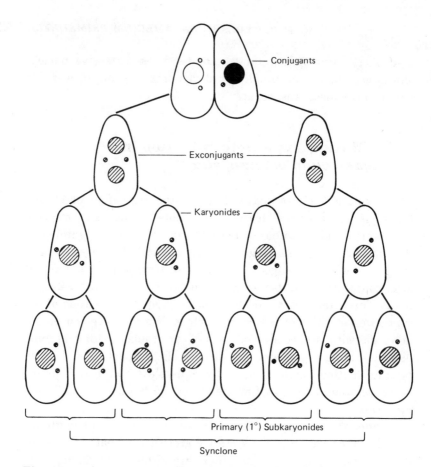

Fig. 13-1. A vegetative pedigree analysis in *Paramecium primaurelia,* whose vegetative cells normally contain one macronucleus and two micronuclei. As a consequence of reciprocal fertilization (Chapter 7), genetically identical new nuclei are established in the two *exconjugants,* whose combined progeny constitute a *synclone.* Each exconjugant characteristically develops two new macronuclei, which are separated at the first cell division after conjugation. The daughter cells of this division are designated as *karyonides,* and all their progeny receive products of an original macronuclear primordium. A vegetative pedigree analysis often involves separating the products of the second postzygotic cell division also. Whenever traits are found to differ between, but not within, karyonides, their basis in heritable differences in the macronuclei must be suspected. (The fragments of the old macronuclei are not included in these drawings.)

226

first division, distributing the two new macronuclei to different daughters, the fission products are again isolated into separate drops of medium where they form the basis of the four karyonidal cultures composing the synclone. The vegetative pedigree may be continued for one or more cell divisions, separating the daughter cells at each division. The daughters at the next division, at which the new macronucleus first divides, give rise to the primary (1°) subkaryonides.

When a set of eight 1° subkaryonides is tested for its mating types, all are found to be of pure type (i.e., they are nonselfing), but usually some are type E and some are type O. However, "sister" primary subkaryonides are characteristically alike. If one 1° subkaryonide is type O, so also is the sister subkaryonide. In contrast, the sister karyonides, produced at the previous cell division, are often different. Sonneborn interpreted this evidence as indicating the assortment of the mating type determinants at the first postconjugal cell division. Since the new macronuclei are the visible structures assorted at this time, he inferred that the macronuclei control the mating type and that diverse macronuclei may develop in the same exconjugant from the same zygote nucleus.

This inference concerning the association between the macronuclei and the mating types was confirmed in a number of ways. When macronuclear regeneration occurs (Chapter 7) no new macronuclei are formed; the macronucleus is restored from pieces of the old macronucleus and the parental mating type is retained. When, in certain strains, more than the usual number of new macronuclei develop, the assortment of mating types may continue as long as new macronuclei are being assorted. Without doubt the karyonide is the unit of mating type determination, and the macronucleus is the organellar locus of the determination.

The karyonidal nature of mating type determination in *P. primaurelia* was further confirmed by Kimball's observations on autogamy—the self-fertilization process occurring in single cells (Chapter 7). The two karyonides produced by a single exautoga-

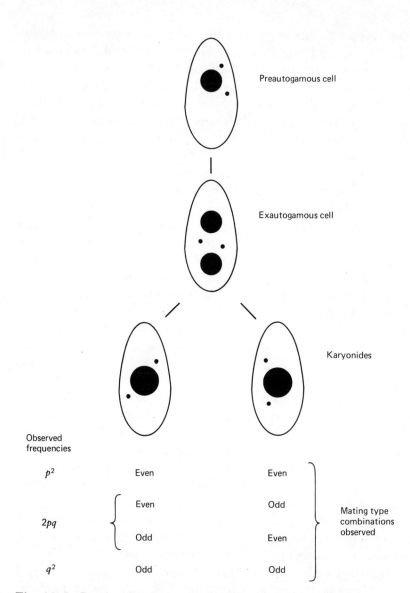

Fig. 13-2. Random karyonidal distribution of mating types in *Paramecium primaurelia* following the process of autogamy (Chapter 7). Autogamy yields a homozygous zygote nucleus that gives rise to two new mac-

mous cell are usually pure for mating type, but one karyonide may manifest type E and the other karyonide type O. The recurring redetermination of mating types at successive autogamies demonstrates that both mating types can be developed on the same homozygous foundation and that the nuclear alterations in one generation do not prejudice those in the next generation.

This question of the relationship between the mating types in successive generations needs to be examined more quantitatively, and to do this we first need to inquire about the developmental behavior of two macronuclei in the same cytoplasm at the same time. We note above that sister karyonides *may* be different, but we do not specify how often they are different. Sonneborn explored this question by allowing a population of cells to undergo autogamy and isolating each karyonide for mating type ascertainment. After classifying the mating types of the karyonides, he then classified the combinations of mating type from different autogamous cells (Fig. 13-2). All possible combinations of sister karyonides were found, and the frequencies of the combinations were those predicted on the basis of chance alone. If, for example, in the entire population of karyonides 60% (p) were type E and 40% (q) were O, then the frequency of autogamous cells producing two type E karyonides was 36% (p^2). Similarly, 16% (q^2) of the autogamous cells gave only type O progeny and 48% ($2pq$) gave one karyonide of each mating type. This result signified the independent determination of the two sister macronuclei.

If two macronuclei developing in the same cytoplasm ignore each other, we would scarcely expect them to be affected by their parent's nuclear condition. Indeed, exautogamous karyonides

←─────────────────────────────────────

ronuclei distributed to daughter cells at the first postzygotic cell division. Karyonides are usually pure for a mating type (all their subkaryonides are alike), but sister karyonides may or may not be alike, depending wholly on chance. If p is the frequency of mating type Even in a large sample of karyonides, and $1 - p = q$ is the frequency of Odd mating types in the same sample, the distribution of mating types conforms to binomial expectations.

produced from a clone of type E have the same frequencies and the same distributions of mating types as do those from a type O clone. When conjugation is induced between a type E and a type O clone, the progeny of the different pair members cannot be distinguished on the basis of their mating types. The distribution of mating types among the four karyonides of the synclones conforms again to binomial expectations: $(p + q)^4$. The frequency of synclones with four type E karyonides is p^4, and so on (Fig. 13-3).

This independence of macronuclear determination does not imply that it is incapable of being influenced by other factors. Indeed, the probabilities of the O–E determinations are markedly sensitive to the temperatures prevailing at the time of conjugation. At low temperatures the O mating type is frequent, but its probability drops as the temperature rises. Even so, the independence of sister karyonides is preserved at all temperatures.

The sensitivity of mating type frequencies to temperature differentials permits temperature to be used as a probe of the time of nuclear differentiation. Early studies of this kind showed that the "sensitive period" is localized during nuclear reorganization rather than, for example, at the time new macronuclei are separated into karyonides, but larger and more detailed analysis might be helpful. So also might be studies to determine if other environmental agents are capable of altering mating type frequencies.

The interpretation arising from the observations reviewed is that mating type determination involves a choice between two alternative nuclear states, that the choice may be influenced by environmental factors during a sensitive period in macronuclear development, and that the choice is perpetuated through the subsequent life of the karyonide. An important insight into the physical nature of the nuclear choice comes from observations on selfers. In some strains essentially all karyonides are pure O or pure E. In other strains (and in certain species), many karyonides self, that is, they contain both mating types and so can conjugate within a culture. Such selfing cultures may be induced in strains with pure karyonides by forcing the fusion of new macronuclei

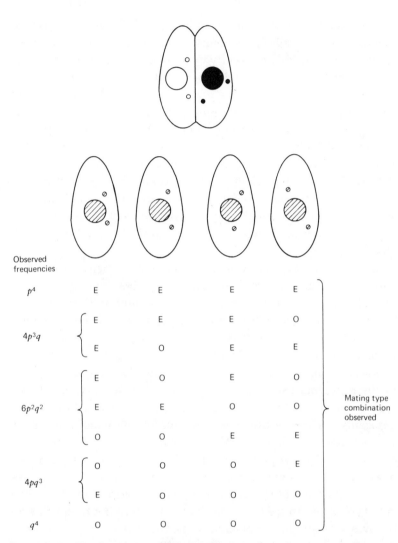

Fig. 13-3. Random karyonidal distribution of mating types in *Paramecium primaurelia* following conjugation. The two mating types (Even and Odd) are distributed at random among the four karyonides of each synclone. If p is the frequency of Even type karyonides and q is the frequency of Odd karyonides, the distribution of mating types within the synclones is predicted by the binomial expansion $(p + q)^4$.

231

through starvation after conjugation, or even by injecting parts of macronuclei into cells of different type. Selfing lines can give rise to stable pure sublines, apparently by a process of assortment, but the assortment has not been studied thoroughly. The determinants for both mating types seem to be able to coexist in the same nuclear envelope and to preserve their integrity, reproduce, and assort. The selfing condition is best explained on the basis of a mixed population of stable elements, rather than an unstable physiological state of the nucleus. This interpretation is based on more substantial data for Tetrahymena than for Paramecium, but the implications are the same. Mating type determination involves a stable, irreversible state associated with some genetic element in the macronucleus. A structural change seems more likely than a physiological change.

Little is known concerning the genetic element involved in determining the mating type specificities, but some genetic differentials are known. Particularly, some strains of *P. primaurelia* never produce karyonides of the E mating type. These are sometimes referred to as *one-type strains* to distinguish them from the more common *two-type strains* we have been discussing. When crosses are made between one-type and two-type strains (Fig. 13–4) the F_1 includes some karyonides of each mating type; the two-type condition is dominant to the one-type condition. At the F_2 by autogamy, as in the previous cases discussed, some autogamous clones contain two E karyonides, some have one of each, and some contain two O karyonides. The distribution in this F_2 is different, however, in that over half of the clones have two O karyonides. When these are studied at a subsequent autogamy, most produce only O progeny. Indeed, the inability to produce the E mating type behaves as a simple Mendelian recessive, *mat-1,* in contrast with the standard dominant *mat-2* allele. The genetic constitution of the developing macronucleus defines its mating type capabilities, but other circumstances determine which potentiality is realized whenever alternatives are possible. One-type stocks have been found in other species of the *P. aurelia* complex also. As in this case the restriction is nearly always to the Odd mating type, though

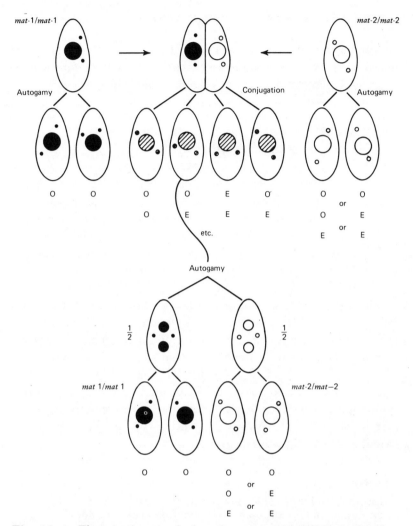

Fig. 13-4. The genetic control of mating type potentialities in *Paramecium primaurelia* illustrated by crosses between a one-type strain only able to produce Odd type progeny and a normal two-type strain. The one-type strain is homozygous for an allele *mat-1* that limits its autogamous progeny to type Odd. The two-type strain has the *mat-2* allele and produces karyonides randomly determined for mating type. The F_1 between the strains has both kinds of karyonides, demonstrating the dominance of the *mat-2* allele. The F_2 by autogamy consists of one half restricted to the Odd mating type and one half showing random karyonidal distribution.

Brygoo has reported some exceptions. The asymmetry of recovered mutations may be a consequence of the structural basis for the mating type choices, but we are not yet able to describe that basis.

C. Mating Type Determination in *Tetrahymena thermophila*

The phenomena of mating type determination in *T. thermophila* are remarkably similar to those in *P. primaurelia,* except that they are more compound. This species has seven mating types instead of two, but most karyonides are pure for one of the seven types. As in *P. primaurelia* the mating type potentialities are controlled by a single known genetic locus, with a limited number of alleles. Karyonides homozygous for *mat-1/mat-1* (formerly *mt*^A) may manifest types I, II, III, V, or VI (Table 13-1). Those homozygous for *mat-2/mat-2* (formerly *mt*^B) may manifest II, III, IV, V, VI, or VII, in characteristic frequencies. Heterozygous *mat-1/mat-2* karyonides may express any of the seven known mating types.

As in the case of *P. primaurelia,* and within the genetic constraints mentioned above, the karyonides produced in the same synkaryon are independently determined as to mating type. No correlation is found between the mating type of the cytoplasmic parent and that of the offspring.

TABLE 13-1

Mating type frequencies among karyonides in three genotypes of *Tetrahymena thermophila* crosses under standard conditions

Genotype	Mating Type Frequencies						
	I	II	III	IV	V	VI	VII
mat-1/mat-1	0.25	0.17	0.17	0.00	0.10	0.30	0.00
mat-1/mat-2	0.14	0.20	0.11	0.23	0.09	0.18	0.06
mat-2/mat-2	0.00	0.15	0.09	0.47	0.05	0.14	0.10

TABLE 13-2

Mating type frequencies among *mat-1/mat-2* **heterozygous karyonides of** *Tetrahymena thermophila* **observed when conjugation occurs at different temperatures**

Temperature (°C)	Mating Type Frequencies						
	I	II	III	IV	V	VI	VII
12	0.07	0.25	0.40	0.08	0.11	0.04	0.06
16	0.12	0.23	0.24	0.13	0.09	0.13	0.06
23	0.14	0.20	0.11	0.23	0.09	0.18	0.06
30	0.16	0.17	0.05	0.26	0.05	0.27	0.04
34	0.09	0.15	0.03	0.49	0.04	0.17	0.03

Again, as in the case of *P. primaurelia,* the frequencies of the mating types are sensitive to the temperature prevailing at the time of conjugation. The frequencies of some mating types (IV, VI) rise with temperature, some fall markedly with temperature (III), and some show relatively slight or complex responses (Table 13-2). The mechanism of mating type determination is not understood, but some attempts have been made to interpret the Tetrahymena system as a compound form of the *P. primaurelia* device. In each case attention is focused on a chromosomal segment, which in the case of *P. primaurelia* contains a bistable element—capable of indefinite persistence in either of two forms once it has passed through a developmental program. To account for seven stable states, the *T. thermophila* system would require at least three bistable components, perhaps homologous in origin or analogous in function to those in Paramecium. These elements, in any case, would have to be sensitive to many kinds of environmental influence, for the proportions of the mating types may be modified in many ways.

One of the contributions of Tetrahymena studies has been in the analysis of mosaic macronuclei. Although most karyonides are pure for mating type, a small percentage are selfers. When selfing karyonides are expanded by making single cell isolations, some sublines are pure for mating type and remain so. Sublines that self may be expanded again, and new pure sublines are obtained. Such studies reveal several significant features of the mosaics.

1. Most selfing karyonides produce only two kinds of pure sublines, though some yield three or four.

2. Any combination of mating types permitted by the genotype may be recovered from some karyonide, though the frequencies of the combinations may be somewhat biased.

3. Most selfing karyonides yield one mating type in great excess and the other mating type only rarely.

4. The assortment rate for pure mating types after equilibration (Chapter 8) is the same for any combination of mating types; in fact, assortment was first studied for mating types, and the 0.0113 fixation rate was first defined for this system.

The properties of the mosaic nuclei and particularly their detailed quantitative behavior provide the chief evidence for discrete structural elements, equivalent in number to the number of haploid sets in the macronucleus, differentiated for genetic function, coexisting in the same nuclear environment.

SUMMARY

Karyonidal determination is a peculiarly ciliate device for generating stable hereditary differences within genetically homogeneous populations. The most thoroughly studied examples involve the fixation of mating type specificity in cells with multiple (or at least dual) potentialities. At a critical stage in macronuclear development following conjugation (or autogamy), the macronuclei undergo an irreversible determination. The two macronuclei developing in the same cytoplasmic environment show no correlation in their choice of mating type, nor are correlations found between the mating type of the cytoplasmic parent and that of its progeny. In each generation, and in each macronucleus, an independent determination fixes the mating type for the rest of the clonal history.

Although the nuclear determinations are probabilistic, the probabilities for the events may be influenced in various ways. Particularly, the frequencies of the mating types are strongly dependent on the temperature at the critical time, and the array of mating type potentialities is controlled by conventionally transmitted chromosomal alleles.

In some species macronuclei arise (or may be constructed) with mixtures of differentiated elements. The manner in which mosaic macronuclei assort the mating type determinants illuminates both the nature of the elements and the organization of the macronucleus.

RECOMMENDED READING

Butzel, H. M., Jr. 1974. Mating type determination and development in *Paramecium aurelia*. In *Paramecium: A Current Survey* (W. J. van Wagtendonk, Ed.), Elsevier, pp. 91–130.

Nanney, D. L. 1956. Caryonidal inheritance and nuclear differentiation. *Am. Nat.*, **90**, 291–307.

Nanney, D. L., E. B. Meyer, and S. S. Chen. 1977. Perturbance analysis of nuclear determination in Tetrahymena. I. Background, rationale, and illustrative example employing temperature responses. *Differentiation*, **9**, 119–130.

Phillips, R. B. 1969. Mating type inheritance in syngen 7 of *Tetrahymena pyriformis*; Intra- and interallelic interactions. *Genetics*, **63**, 349–359.

Sonneborn, T. M. 1977. Genetics of cellular differentiation: Stable nuclear differentiation in eucaryotic unicells. *Annu. Rev. Genet.*, **11**, 349–367.

Coordinated Karyonidal Determination and Related Phenomena

14

A. Mating Type Determination in
Paramecium tetraurelia

The mating type system in *P. tetraurelia* is believed to be a variant of that in *P. primaurelia,* although it has some interesting new features. The reasons for associating the two systems include the evidence of homology among the mating types. Some interspecific matings occur between species with systems like that in *P. primaurelia* (the Group A species) and species with systems like that in *P. tetraurelia* (the Group B species). Although these matings are sterile, leading to inviability in either F_1 or F_2, they are sufficient to establish the homologies among the Even mating types and among the Odd mating types of most of the species.

In spite of the similarities in the end products of mating type determination, the methods of mating type determination appear at first to be very different. In *P. tetraurelia,* and the Group B species generally, the two karyonides from an exconjugant (or exautogamont) are ordinarily of the same mating type and are of the same mating type as the cytoplasmic parent (Fig. 14-1). Thus when conjugation occurs between an Even and an Odd cell, one exconjugant clone is usually Even and the other is Odd. The mating types appear to be associated not with the developing macronuclei, but with the cytoplasm.

This implication of the cytoplasm in mating type determination was reinforced by studies using cytoplasmic exchange (Chapter 7). When extensive cytoplasmic exchange is induced during conjugation (Fig. 14-2), the two exconjugant clones are usually of the same mating type. A central issue then became the reconciliation of nuclear determination of mating types in some species and apparent cytoplasmic determination of similar specificities in other species.

The reconciliation of these two patterns came through the proposal that both involved nuclear determination, but that one (the Group B) also involved an induction of that nuclear determination

241

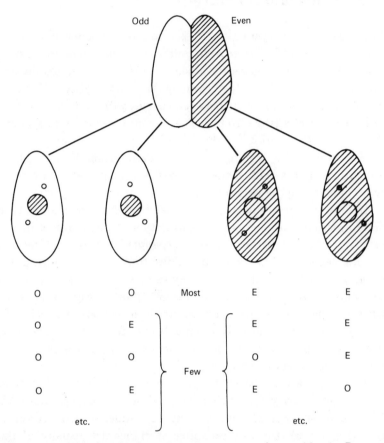

Fig. 14-1. The distribution of mating types to karyonides in *Parame-cium tetraurelia* that manifests coordinated karyonidal distribution. The two karyonides from the original type Odd cytoplasmic parent are usually Odd. The karyonides from the Even parent are usually Even. However, Odd karyonides from Even parents and Even karyonides from Odd parents are sometimes observed.

242

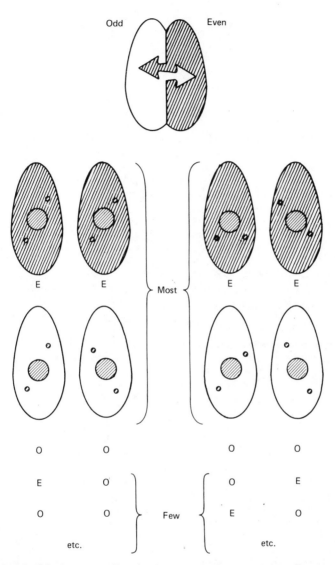

Fig. 14-2. Mating type distribution among karyonides in *Paramecium tetraurelia* following extensive cytoplasmic exchange at conjugation. Most synclones (sets of four karyonides) are pure for mating type, but again sister karyonides are occasionally dissimilar.

243

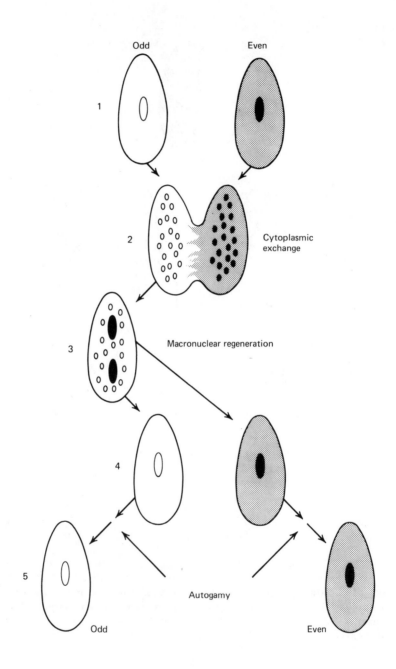

Even

1

2

Cytoplasmic
exchange

3

Macronuclear regeneration

4

5

Autogamy

Odd

Even

244

through a cytoplasmic influence. The cytoplasmic influence was in turn believed to be controlled by the old, already determined macronucleus that coexisted in the cytoplasm with the developing macronuclear primordia.

This interpretation was subjected to critical experimental tests by Sonneborn. He first attempted to establish cells containing macronuclei determined to express different mating types (Fig. 14–3). This objective was approached as follows. When cytoplasmic exchange is induced (by antiserum treatment), both exconjugants ordinarily yield clones of the same mating type. At higher temperatures both clones are usually the E type. According to the hypothesis just outlined, the new macronuclei in one of the cells should be different from the old macronucleus, which has fragmented. If the two kinds of macronuclei are the foci of control, then their assortment should coincide with mating type assortment in vegetative pedigrees. The assortment can be arranged by the process of macronuclear regeneration (Chapter 7). High temperature at a critical time in their development retards the macronuclear primordia and permits the fragments of the old macronucleus to persist. For several cell divisions the new macronuclei may be carried along without dividing before they recover from the temperature treatment. At each cell division one daughter cell receives the primordium (and some fragments), and the other daughter receives only fragments. When the primordium begins to divide, any residual fragments are repressed and eventually

Fig. 14–3. Analysis of mating type determination in *Paramecium tetraurelia* through cytoplasmic exchange and macronuclear regeneration. (*1*) Cells of complementary mating type. (*2*) Cells conjugate and macronuclei fragment; cytoplasmic exchange is induced. (*3*) Under these conditions new macronuclei in both exconjugants are type Even, but in the former Odd type cell the macronuclear fragments are from an Even type macronucleus. (*4*) Macronuclear regeneration and assortment is induced by temperature shocks, yielding some Even and some Odd sublines. (*5*) At a subsequent autogamy (or conjugation) the cells breed true for mating type.

resorbed. When a primordium is lost, by assortment, the fragments are released from inhibition, grow, and are assorted to become the definitive macronuclei of the cell. These events may be monitored by cytological preparations of some of the cells or by means of genetic markers for the two kinds of macronuclei. The results of the experiments are unequivocal. When a cell lineage is established from the new primordium, its mating type is E. When a lineage is established with fragments from an O type cell, that lineage is O. The macronuclei control the mating types as clearly in the Group B species as in the Group A species.

Thus far unresolved, however, is the source of the cytoplasmic influence on macronuclear determination. As mentioned earlier the macronucleus might be determined by the cytoplasm, but the cytoplasm is in turn controlled by the macronucleus–in a cyclic nucleo-cytoplasmic interaction. The alternative is that the cytoplasmic condition might be self-perpetuating and indifferent to the presence of determined macronuclei. A test of these two hypotheses was provided by further study of the assorting clones just described. Experiments have been carried on in which cytoplasm is mixed under circumstances (high temperature) appropriate for the predominance of the E type influence. Within this mixed cytoplasm are maintained for awhile two kinds of macronuclei, but after a few cell divisions the macronuclei and the mating types are allowed to assort. If the cytoplasmic conditions are not coupled to the nuclear conditions, all lineages should have E type cytoplasm (assuming complete exclusion of O determinants), or else some lineages should have E and some O determinants, with no necessary relationship to the kind of macronucleus present.

The kind of cytoplasm present can be assessed by allowing another nuclear reorganization (autogamy) in these lines. When autogamy occurs the lines with E mating types produce E progeny, and the lines with O mating types produce O progeny. Thus the cytoplasmic conditions for the induction of mating type assort at precisely the same cell divisions at which the new and old macronuclei assort. The cytoplasmic state must be a function of the macronucleus present in the cell. The determined macronucleus produces a substance or promotes a condition capable of inducing the same macronuclear determination in a naive macronucleus developing in its vicinity. Perhaps the chief difference between the Group A and Group B species lies in this ability of macronuclei of Group B to influence determination at a distance. Such influences may occur in Group A also, but only among the genomes within a single nuclear envelope.

B. Other Mating Type Systems in Ciliates

The ciliated protozoa provide a remarkable array of mating type phenomena, and no comprehensive summary is possible within this book. We comment earlier on differences in the numbers of mating types in different species (Chapter 6), from two in the *P. aurelia* complex to dozens in some of the hypotrichs. Some species in fact have no known mating types and have never been observed to mate. The amicronucleate ciliates, such as *T. pyriformis (strictu senso)* have abandoned sexuality entirely and must rely on somatic mutation and assortment for evolutionary variety. Other ciliates, such as *T. rostrata*, are also not known to conjugate, but they do have micronuclei, and they periodically return to the germinal nucleus in a process of autogamy. Finally, some ciliates are regularly observed to conjugate in clonal cultures, but pure mating types have not been established; cells within the species may have transient mating types or may be able to conjugate without restrictions.

The ciliates about which we know the most, however, are those capable of maintaining pure mating types in culture, because these are the ones suitable for controlled breeding studies. These ciliates fall into two major classes with respect to mating type determination: those in which all the progeny of a conjugating pair (a synclone) are alike in mating type, and those in which diversities are regularly present within a synclone. Synclone uniformity is the result expected for direct genotypic control of mating types, and systems with synclonal uniformity are sometimes characterized as "genetic," as opposed to those "epigenetic" systems in which synclones are of mixed mating types. However, the notion that "karyonidal" determination involves only developmental or regulatory differences between the mating types is questionable. We argue earlier (Chapter 13) that "mosaic" macronuclei seem to require structural, that is, genetic, differentials among macronuclear elements. The nuclear differences arising regularly as part of a developmental program may therefore be true genetic differences, and the so-called genetic and epigenetic mating type systems may be more alike than they originally appeared to be.

The genetic systems are themselves highly diverse, and we can only briefly review some of them here. The species complex of *P. bursaria* provides one well-studied example. Each genetic species contains either four or eight mating types. Mating type determination is by either a two locus or a three locus mechanism. Each locus is represented by two alleles, a dominant and a recessive. A two locus system then makes possible four classes of genotypes: *A-B-, A-bb, aaB-* and *aabb,* corresponding to the four mating types. A three locus system is comparable, but potentiates eight classes of genotypes and eight mating types. The major features of the *P. bursaria* systems are accounted for by these simple genetic considerations. Under special circumstances, however, new features appear that suggest at least some additional regulatory elements. We mention earlier (Chapter 6) the phenomenon of "senescent" selfing in *P. bursaria* in which mating types in old clones sometimes become "destabilized" and thus make possible the ordinarily forbidden intraclonal conjugation. We should also mention the peculiar interactions observed occasionally in abortive conjugation that may alter permanently the mating behavior of a cell line. Both these phenomena suggest that the genetic states normally transmitted through classic meiotic processes have similarities to the genetic states peculiarly susceptible to developmental influences in other ciliates.

Another example of a synclonal system is that of *Tetrahymena pigmentosa*. Orias showed that this species has three mating types controlled by alleles at one locus. The alleles manifest "peck order" dominance. *Mat-1* is dominant over *mat-2* and *mat-3* and determines mating type I. *Mat-2* is dominant over *mat-3* and is associated with type II. The homozygous *mat-3/mat-3* expresses type III. Such peck order series are not uncommon in the ciliates. The unconventionality in this system is found in the instability of the *mat* alleles. Both *mat-1* and *mat-2* show a high frequency of "mutation" to the *mat-3* state; the frequency is about 5% per generation. The allelic changes do not occur only in the micronucleus, but are probably also responsible for unstable or selfing karyonides. Again, we are led to suspect a relationship between these micronuclear mutations and the macronuclear states established in other Tetrahymena species by macronuclear determination.

Still another genetic system with interesting instabilities is that of *Euplotes crassus*, a hypotrich ciliate very different from Tetrahymena, but again showing a peck order series of alleles controlling as many different mating types. Heckmann found that heterozygotes at about 400 cell divisions began regularly to express their two alleles alternately in cell lineages, even though the more "recessive" allele was unexpressed during the previous hundreds of cell divisions. This alternate expression allows conjugation within a clone, a practice prohibited in younger clones (Chapter 6), but permissible in old clones. In the present context, however, the senescent selfing may be rationalized as just another example of regulatory intervention in the expression of genetically determined mating type characteristics.

C. Karyonidal Determination of Traits Other Than Mating Types

Although nuclear differentiation has been extensively exploited by the ciliates for the regulation of mating phenomena, karyonidal distribution is not limited to mating types. Moreover, many cases of nuclear differentiation may be missed simply because all the developing nuclei are responding to similar cytoplasmic or environmental signals. Karyonidal determination is most apparent when chance plays a sufficient role to permit differences to be expressed in sister karyonides.

One interesting example of karyonidal determination of a different kind of trait is provided by Genermont's study of calcium chloride resistance in *P. aurelia*. This careful quantitative analysis demonstrated both a karyonidal distribution of salt resistance and also a cytoplasmic effect on the level of resistance.

The most recent studies are those of Sonneborn and his colleagues on yet a different kind of trait—the ability to discharge the cortical organelles called trichocysts. The ability versus nonability to discharge trichocysts is inherited at conjugation in *P. tetraurelia* in a manner very similar to that of mating types in this species. A closed cycle of nucleo-cytoplasmic interactions has been demonstrated; occasional karyonidal differences have been observed and can be regularly induced; environmental influences on the probabilities of the nuclear states can be shown.

All these examples demonstrate stable functional differences among somatic nuclei, established within genotypically determined limits at a particular developmental stage. Because many of the differences are effectively irreversible once they are established, these nuclear alterations are probably of a different type than the more readily reversible changes observed with serotypes (Chapter 12). An unequivocal means of distinguishing between functional inertia in the nuclear apparatus and structural differences among originally identical nuclei is not available yet, but grounds certainly exist for suspecting that regulated genetic alterations form a significant part of the tactical repertoire of the ciliates.

SUMMARY

A significant elaboration of the process of nuclear determination brings it under control by making it responsive to signals from other (parental) determined nuclei. Particularly, in some species the two karyonides from one exconjugant are characteristically of one mating type and those from the other bear the mating type of the other parent. A role for the cytoplasm is demonstrated by the effects of cytoplasmic fusion; the four karyonides are usually of the same mating type. The role of the macronucleus is shown by the occasional occurrence of unlike sister karyonides, but more impressively by controlled experiments in which the vegetative assortment of macronuclear fragments and new macronuclear primordia is followed. The system may be viewed as a closed cycle of nucleocytoplasmic interaction. Signal coordinated karyonidal determination mimics "cytoplasmic inheritance," because indelible information may be transmitted through cytoplasmic connections. It differs from "true" cytoplasmic inheritance in that the replication of the relevant differences in the present instance is nuclear.

Coordinated nuclear determination is probably much more common among ciliates than is random nuclear determination, but its basis would probably be entirely obscure were it not for the hints provided by the random mode.

RECOMMENDED READING

Barnett, A. 1966. A circadian rhythm of mating type reversals in *Paramecium multimicronucleatum*, syngen 2, and its genetic control. *J. Cellular Comp. Physiol.*, **67**, 239–270.

Brygoo, Y. 1977. Genetic analysis of mating-type differentiation in *Paramecium tetraurelia. Genetics,* **87**, 633–653.

Génermont, J. 1961. Déterminants génétiques macronucléaires et cytoplasmique controlant la résistance au chlorure de calcium chez *Paramecium aurelia* (souche 90; variété 1). *Ann. Genet.,* **3**, 1–8.

Hiwatashi, K. 1968. Determination and inheritance of mating type in *Paramecium caudatum. Genetics,* **58**, 373–386.

Nobili, R. 1966. Mating types and mating type inheritance in *Euplotes minuta* Yocum (Ciliata, Hypotrichida). *J. Protozool.,* **13**, 38–41.

Orias, E. 1963. Mating type determination in variety 8, *Tetrahymena pyriformis*. *Genetics,* **48,** 1509–1518.

Sonneborn, T. M. 1954. Patterns of nucleocytoplasmic integration in Paramecium. *Caryologia,* **6** (Suppl.), 307–325.

Sonneborn, T. M. and M. V. Schneller. 1979. A genetic system for alternative stable characteristics in genomically identical homozygous clones. *Develop. Genetics,* **1,** 21–46.

15

Cytoplasmic Inheritance and Intracellular Symbionts

A. Criteria for Answering the Geographical Question

The subject of cytoplasmic inheritance has long been associated with the ciliated protozoa, primarily because several examples of unconventional transmission of traits were described in ciliates shortly after the discovery of mating types permitted such descriptions. We need to understand why so many examples of unconventional transmission appeared in these organisms, and we also need to consider the basis for labeling the phenomena "cytoplasmic."

As we remark earlier (Chapter 7), the ciliate life cycle is distinctive in (among other things) its provisions to transmit to a single zygote the entire organismic protoplasm of a parent. Although the parent may be called a "cell," according to a permissive definition, it is compound in sustaining many copies of parts of the genome—dozens or even hundreds. Yet this somatic mass is passed entirely to a single diploid zygote.

Not only should we be impressed with the sheer weight of a ciliate's somatic heritage, but we should also notice its state of organization. Multicellular organisms may or may not contribute to their progeny large amounts of protoplasm, but when a substantial investment is made, it is usually converted into bland nutrients in a cryptic organization and is not transmitted as fully differentiated adult equipment. Yet the ciliate's inheritance involves a fully functioning and integrated system of organelles, of great complexity, which is only gradually taken over by a new nucleus and shaped to new specifications in the course of several protoplasmic doublings (*phenomic lag*).

Under these circumstances one should not be surprised at the abundance of evidence available on ciliates for the transmission of information between generations by somewhat unconventional methods. Such an understanding need not imply that these unconventional methods are unique to the ciliates and hence of limited interest to students of other life forms. Rather, one should suspect that the methods are universal, but that their employment in intergenerational transactions is somewhat unusual. The usual arena of their employment, particularly in metazoans, might well be the maintenance of somatic specializations.

The ciliates have not only provided several interesting examples of "cytoplasmic heredity", but a methodology uniquely suited to their characterization. Particularly, every conjugating pair consists of precisely controlled reciprocal products; one exconjugant receives the somatic investment of one parental lineage and the other receives a different somatic contribution, but the zygote nuclei are genetically identical. Any parental distinctions that persist through conjugation are automatically recognized as being in some way "cytoplasmic." The inference of a role of the cytoplasm in the continuity of specificities is often confirmed by the controlled mixing of cytoplasm during conjugation. Distinctions that remain when only nuclei are exchanged and that disappear when cytoplasms become confluent are confirmed instances of cytoplasmic inheritance.

One might expect that all the cases of cytoplasmic inheritance in ciliates might prove to be mechanistically similar, but by no means has this been found to be true. Indeed, each case of cytoplasmic inheritance seemed for awhile to be distinctive in important respects. In recent years, however, it has been possible to discern a limited number of categories of cytoplasmic information transfer. We can illustrate these categories by a cross, which can be carried out in *Paramecium tetraurelia.*

Let us begin with a pair of strains that differ in five characteristics: mating type, antigenic type, chloramphenicol sensitivity, cell morphology, and the killer trait. Strain 1 is mating type Odd, antigenic type A, chloramphenicol sensitive, a doublet, and a killer. Strain 2 is mating type Even, antigenic type B, chloramphenicol resistant, a singlet, and sensitive to the killer toxin. Using other hereditary differences with more conventional transmission patterns, we can confirm that the cross is completed normally and that the two exconjugant clones are alike in their nuclear genes.

Yet, when we examine the clones for mating types, serotypes, drug resistance, cell morphology, and the killer trait, they show the same distinctions as did the strains from which they were derived; the nuclear exchange appears inconsequential.

As a next step in the analysis we induce conjugating pairs to undergo massive cytoplasmic exchange, as well as to exchange pronuclei. The results are very different. Both exconjugant clones are now of the same mating type (usually Even), both are the same antigenic type (A or B), both are drug resistant (if grown in its presence), and both are killers. Only with respect to cell morphology do the preconjugal distinctions persist; the doublet exconjugant yields a doublet clone and the singlet exconjugant produces a singlet clone. The disappearance of the parental distinctions in the presence of cytoplasmic mixing confirms a significant role of the cytoplasm in the maintenance of the distinctions. But the failure of the two exconjugant clones to become alike with respect to the cortical morphology also indicates a role of the cytoplasm. This result in fact makes it possible to distinguish between two parts of the cytoplasm: the solated cytoplasm easily exchanged through conjugal bridges, and the gelated cortex that retains its integrity.

The operation of "reciprocal crosses" makes it possible to determine cytoplasmic roles and even to probe the regions of the cytoplasm responsible for continuity of specificity. The results do not, however, in either case, demonstrate an *exclusively* cytoplasmic basis for information transfer. That is, we may not *a priori* conclude that cellular homeostasis in any of these cases is maintained solely by organelles in one of the two geographical regions defined by the nuclear membrane. We cannot say that chromosomal genes are responsible in one case and cytoplasmic genes in another.

We discuss in previous chapters examples of both structural and functional inertia, cases of persistent cellular differences maintained in the presence of identical genetic components. The morphological distinctions between singlets and doublets (Chapter

10) probably involve no molecular differences whatever; the distinctions between lineages expressing different nonallelic serotypes (Chapter 12) certainly involve molecular differences, but they probably do not involve differences in the structures of the genes, only in their functional states, maintained through a physiological system of positive feedback.

When we come to mating type differences in this species (Chapter 14), we have a provocative example of the hazards of the operational distinctions used in separating nuclear and cytoplasmic heredity. The effects of reciprocal crosses and of cytoplasmic mixing seem to indicate a "simple" case of cytoplasmic determination. That judgment was first called into question, however, because of the method of mating type determination in closely related species. In the so-called Group A species, sister macronuclei developing in the same cytoplasm are independently determined as to mating type and maintain that determination as long as the life cycle continues (Chapter 13). How is the cytoplasmic inheritance of mating types in one species reconciled with the nuclear basis of homologous specifications in another species?

The resolution of this enigma hinged on the recognition that the test for cytoplasmic inheritance established merely the existence of *a* cytoplasmic component in the hereditary apparatus, not of an exclusively cytoplasmic system. In fact, both the Group A (karyonidal) species and the Group B (cytoplasmic) species have nuclear mating type determination. The difference lies in the ability of the B species to transfer information concerning the state of nuclear differentiation through cytoplasmic connections. An Even type nucleus "instructs" macronuclear primordia by way of cytoplasmic messages to become determined in the same way it had been determined. Experiments by Sonneborn showed that the cytoplasm plays no independent role in this information transfer, but only serves as a passive route of communication. The nature of the nuclear change signaled through the cytoplasm is not certainly established. It is stable enough to suggest that it involves a true structural alteration of the chromatinic materials. Yet, in the initial phases of the study, mating types in the species of the B group of the *P. aurelia* complex (e.g., *P. tetraurelia*) were thought to be cytoplasmically inherited.

B. Native and Foreign Genetic Elements

The experiences mentioned above with cortical structures, serotype expression, and mating type determination indicate a need for caution in diagnosing cytoplasmic inheritance on the basis of reciprocal crosses, however adequate this procedure may be in other organisms. All these examples involve complex nucleocytoplasmic interactions with diverse roles for the cytoplasm. In other cases, however, the same preliminary methodologies have led to the identification of cytoplasmic components with unequivocal genetic continuity, and particularly components bearing mutable nucleic reservoirs of their own. The existence of such cytoplasmic genetic depots—in plastids and mitochondria particularly—is now well established, but at the beginning of these studies in Paramecium precedents were virtually unknown outside the higher plants. The first fully examined case of a cytoplasmic genetic system happened not to involve either of these organelles, but more complex structures now conceded to be bacteria.

These bacterial symbionts are conventionally designated by Greek letters—*lambda, kappa,* and so on. Their analysis is probably the best publicized of ciliate studies. Paramecia that release a substance toxic to other paramecia are called killers; those that die when exposed to the toxin are called sensitives. Killers may be crossed to sensitives because under certain conditions sensitives are immune to the toxin; particularly, mating cells are not killed. When ordinary crosses are made between killers and sensitives, one exconjugant clone remains a killer clone and the other remains sensitive, as is noted above (Fig. 15-1). And when cytoplasmic exchange occurs, both exconjugant clones are killers.

Killers may be transformed into sensitives in various ways. The symbionts may be outgrown by culturing the paramecia at ele-

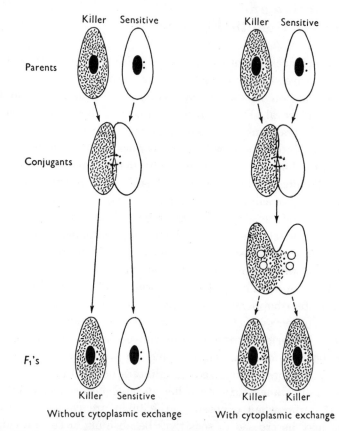

Fig. 15-1. Conjugation between killer and sensitive strains of *Paramecium tetraurelia* with (right) and without (left) cytoplasmic exchange. The nuclear events are identical, but the genetic results differ. Reprinted, by permission, from Beale, G. H. 1954. *Genetics of* Paramecium aurelia, Cambridge University Press. Fig. 3, p. 53.

vated temperatures. They may be destroyed by antibiotics and they may be inactivated by X-rays. In fact, Preer's radiation studies provided the first evidence that the structures responsible for killing ability were large and numerous. Paramecium, because of its genetic compoundness, is highly resistant to radiation damage; this insensitivity of the host permits sophisticated analysis of the symbionts' radiation responses. Preer discovered that the loss of killing ability with X-rays followed single hit kinetics. The slope of the inactivation curve was steep, indicating a large radiation target. The extrapolation of the linear portion of the inactivation curve to the origin suggested that each killer cell might have several hundred large genetic elements in its cytoplasm. Staining the cells with DNA specific stains revealed the visible kappa particles predicted on indirect evidence.

The killer trait is not a "simple" case of infection, or even of a primitive venereal disease. The interactions between the host and the symbiont are complex and mutual. The presence of the kappa particles is almost certainly a burden on the paramecium, though evidence for the cost of maintenance is meager. The symbiont probably confers selective advantages upon the host, but the circumstances in nature where this is true are not known. Paramecia can grow well in laboratory culture without symbionts, but the symbionts can be grown only with difficulty or not at all when separated from their hosts. With other ciliates, such as many freshwater species of Euplotes, the mutual dependence is more striking; the bacterial symbionts (quite different from those in killer Paramecium) cannot survive alone, and neither can the ciliates.

Even in Paramecium the coadaptation of the nuclear and cytoplasmic genetic systems may be noted. Kappa particles do not survive in all Paramecium genotypes, for example. One major gene, the K gene was early identified as an essential factor for kappa maintenance (Fig. 15-2). A homozygous KK paramecium might be phenotypically sensitive, but it is capable of being infected—by cytoplasmic exchange, by microinjection, or occasionally by

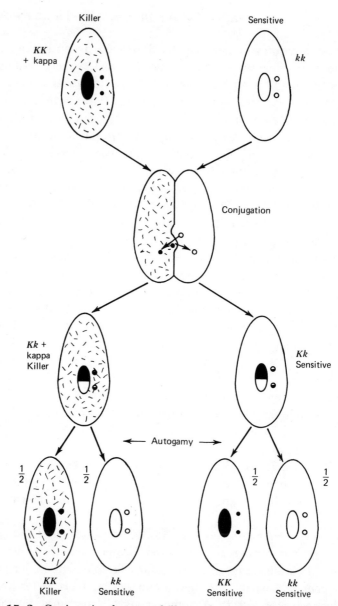

Fig. 15-2. Conjugation between killer and sensitive strains of *Parame-cium tetraurelia* that differ with respect to both the presence of kappa and the presence of the dominant gene *K* necessary for the maintenance of kappa. The F$_1$'s (without cytoplasmic exchange) are killers and sensitives respectively, as in 15-1. However, half the autogamous progeny of the killer F$_1$'s become sensitive, after they lose the *K* gene.

external exposure—under special conditions. In contrast, a *kk* paramecium, even if it contains kappa particles (as when a *Kk* killer becomes homozygous after autogamy), cannot persist as a killer. The kappa particles are diluted out. The kinetics of loss of kappa are curious and ill-understood, but pose no major difficulties in the analysis.

Cytoplasmic symbionts of diverse nature and effect are known among the ciliates. Some particles are associated with no killing action of any kind (some mutants of kappa, for example); some kill only during the mating process (unlike kappa) and are hence known as "mate killers." Most of the symbionts are bacteria, but they seem to have come from several sources and to have undergone a long evolutionary adjustment to their hosts. Some symbionts, however, are not bacteria, but green algae. All strains of *P. bursaria* collected in nature are green, because of the several hundred chlorellae (green algae) contained in their cytoplasm. *P. bursaria* may be "bleached" by growing them in the dark for a long period of time, but the chlorellae must contribute some useful functions to their hosts to explain their ubiquity in natural strains.

Many special problems have been raised by the ciliate symbionts and their interrelations with their hosts, and these problems continue to be studied. One may not, however, question the level of organization represented by the symbionts. They are bona fide organisms of either prokaryotic or eukaryotic grade, showing clear affinities to autonomous organisms, and they are in some cases capable of independent growth when separated from their hosts.

C. Organellar Genetics

The ciliates are not only useful organisms in which to study symbiosis among various kinds of organisms, but they are also suitable for the study of those subcellular structures readily accepted as organelles. We discuss earlier (Chapter 10) at some length the continuity and organization of the basal bodies, because persistent differences with respect to these structures are maintained by structural inertia at a supramolecular level, rather than by conservative (i.e., semiconservative) replication of differences in macromolecules. More conventional mechanisms are involved with the mitochondria.

As with other eukaryotes (yeasts and chlamydomonads), mitochondrial mutants may be selected in the ciliates because of the differential sensitivities of mitochondrial and cytoplasmic synthetic systems to various drugs. Because of their large size, and the well-developed systems of cytoplasmic exchange, the ciliates are particularly well suited to certain kinds of experimental manipulations of these cytoplasmic organelles. Crosses can be made, for example, between strains of Paramecium resistant and sensitive to drugs such as chloramphenicol, erythromycin, or mikamycin. As expected of mitochondrial mutants, the drug resistance follows the cytoplasm. With normal crosses one exconjugant is resistant, the other sensitive; with cytoplasmic mixing, both exconjugants become resistant. Such crosses establish a role for a cytoplasmic element in the maintenance of drug resistance, but do not demonstrate that the cytoplasmic element is the mitochondrion.

Evidence on this point comes from microinjection studies. Drug resistant cells may be ruptured and fractionated; the mitochondria may be purified and injected into drug sensitive paramecia. The development of drug resistance in the injected cells constitutes convincing proof that differences in the mitochondria are the basis for the hereditary differences in drug resistance.

The suitability of these large cells for microinjection even permits exploration of species differences in mitochondria. Mitochondria from drug resistant strains of *P. primaurelia* have been injected into *P. septaurelia* and under suitable selective pressure may replace the native mitochondria (Fig. 15-3). Interestingly enough, the reciprocal transfer—even for these closely related species—is not successful, and the transferred mitochondria of *P. primaurelia*, after accommodation in the other

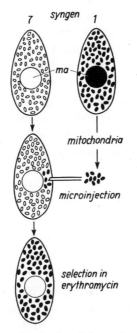

Fig. 15-3. The transfer of mitochondria and associated erythromycin sensitivity from *Paramecium primaurelia* to *Paramecium septaurelia*. Reprinted, by permission from Grell, K. G. 1973. *Protozoology*, Springer-Verlag. Fig. 234, p. 266.

species, cannot be returned to the species of origin. Mitochondria contain many nucleus specified components, and these apparently confer species incompatibility in some combinations.

Attempts have been made to study mitochondrial recombination, and suitably mixed populations have been constructed. Thus far, however, no evidence has been found for recombination of markers carried in separate mitochondria, even though such evidence is available for other unicells.

The mitochondrial DNA in the ciliates thus far studied differs from that in other organisms in its form. Most mitochondrial DNA is circular, though the length of the DNA strand is highly variable in different organisms. In both Paramecium and Tetrahymena the mitochondrial DNA is linear and about 14–15 micrometers in length. It also appears to replicate in an unusual fashion, and in slightly different ways in the two ciliates examined.

One curious and unexpected feature of the mitochondrial DNAs in Tetrahymena is their great divergence in different species of the *T. pyriformis* complex. Different species may show completely different patterns when examined with restriction enzymes, and cross-annealing may occur at only 5–10% of the homologous combinations. These observations are consistent with a great evolutionary plasticity of the mitochondrial DNA and/or an unexpectedly large molecular distance between phenetically similar species (Chapter 5).

Another distinctive feature of the ciliate mitochondrial system concerns their ribosomes. The ribosomes in the mitochondria of ciliates are about the same size as those in the cytoplasm of other eukaryotes (80 S), and hence larger than those in the prokaryotes (70 S). However, unlike either of these other ribosomal types, the ciliate mitochondrial ribosomes separate into subunits of equal size (55 S).

All these features of the ciliate mitochondria suggest a different evolutionary history, and possibly a different evolutionary origin, than other information processing systems in eukaryotes. Intensive comparative studies may be well rewarded.

SUMMARY

Cytoplasmic inheritance based on nucleic templates in the cytoplasm has long been recognized in the ciliates. Examples include the clearly symbiotic associations of green algae with *P. bursaria,* in which both components are capable of independent reproduction. A more intimate association is that with the bacteria responsible for the killer character in the *P. aurelia* complex; at least the bacterium has difficulty becoming established as an independent organism, and the genetic composition of the host provides significant constraints. The hereditary differences associated with mitochondrial DNA reflect an indissoluble union, which may have been achieved through a progressively more intimate interaction of genetic systems of different origin.

RECOMMENDED READING

Adoutte, A., J. Beisson. 1972. Evolution of mixed populations of genetically different mitochondria in *Paramecium aurelia. Nature,* **235,** 393–396.

Ball, G. B. 1958. Organisms living on and in protozoa. *Res. Protozool.,* **3,** 565–718.

Beale, G. H., A. Jurand, and J. R. Preer, Jr. 1969. The classes of endosymbiont of *Paramecium aurelia. J. Cell Sci.,* **5,** 65–91.

Beale, G. H. and J. Knowles. 1978. *Extranuclear Genetics,* Edward Arnold.

Cummings, D. J. 1977. Evidence for semi-conservative replication of mitochondrial DNA from *Paramecium aurelia. J. Mol. Biol.,* **117,** 273–277.

Gillham, N. 1978. *Organelle Heredity,* Raven.

Heckmann, K. 1975. Omicron, ein essentieller Endosymbiont von *Euplotes aediculatus. J. Protozool.,* **22,** 97–104.

Knowles, J. K. C. and A. Tait. 1972. A new method for studying the genetic control of specific mitochondrial proteins in *Paramecium aurelia. Mol. Gen. Genet.,* **117,** 53–59.

Koizumi, S. 1974. Microinjection and transfer of cytoplasm in Paramecium. *Exp. Cell Res.,* **88,** 74–78.

Preer, J. R., L. B. Preer, and A. Jurand. 1974. Kappa and other endosymbionts in *Paramecium aurelia. Bacteriol. Rev.,* **38,** 113–163.

Preer, L. B. and J. R. Preer, Jr. 1977. Inheritance of infectious elements. *Cell Biology: A Comprehensive Treatise* (L. Goldstein and D. M. Prescott, Eds.), Vol. 1 Academic, pp. 319–373.

Sainsard, A. 1976. Gene controlled selection of mitochondria in Paramecium. *Mol. Gen. Genet.,* **145,** 23–30.

Soldo, A. T. 1974. Intracellular particles in Paramecium. In *Paramecium: A Current Survey* (W. J. van Wagtendonk, Ed.), Elsevier, pp. 377–422.

Sonneborn, T. M. 1959. Kappa and related particles in Paramecium. *Adv. Virus Res.,* **6,** 229–356.

Genetic Dissection
of
Cell Structures
and
Functions

16

THE TECHNIQUE OF GENETIC DISSECTION IS ONE OF the most powerful techniques in modern biology. Almost any molecular component of a cell may be modified by mutation, and all representatives of that molecule are modified in precisely the same way. The kinds of modification are far less limited than the investigator's chemical store, or even his imagination; and genetic manipulations permit the molecular components to be added or subtracted, combined, or separated by design.

However, the technique of genetic dissection requires a genetically domesticated organism and a culture regimen suitable to the application of biochemical analysis so that the genetic changes can be interpreted in molecular terms.

As discussed in Chapter 1, a proper research strategy must direct analytic effort to an organism evolutionarily placed so as to possess the biological mechanisms one wishes to examine. The organism should also be as simple a representative of that evolutionary grade as possible. Much of the past effort in ciliate studies is based on these considerations. Organisms are no less research technologies than are mechanical or electronic contrivances, and biologists need new organismic techniques at the eukaryotic level, preferably microbial. Ciliates, along with a few other organisms, are being recommended to this role. Their ability to fulfill its demanding terms is yet unproven. They are only now approaching a state of domestication in which their capabilities can be fully tried. Our remaining task is to summarize the rationales and preliminary results of a few of these efforts.

A. Electrophysiological Functions of Membranes

One of the most ambitious and promising of the genetic dissection efforts is that of Ching Kung and his associates at the University of Wisconsin, using behavioral mutants in Paramecium. The

271

behavior of paramecia has been studied since the pioneer work of H. S. Jennings at the turn of the century. Particularly well known is the "avoiding reaction" (Fig. 16-1) whereby cells meeting a barrier or a toxic substance stop, back up, change direction, and move away. Kung has undertaken a genetic dissection of this behavior in an attempt to relate it meaningfully to larger issues of membrane structure and electrophysiological function.

In spite of many advances in nerve physiology, the functions of

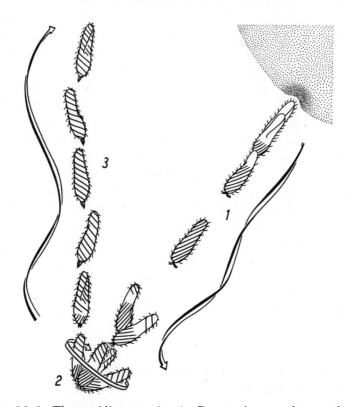

Fig. 16-1. The avoiding reaction in *Paramecium caudatum,* showing changes in ciliary beat associated with (*1*) backward motion after stimulation, (*2*) cone swinging phase, and (*3*) forward motion. Reprinted, by permission, from Grell, K. G. 1973. *Protozoology,* (Springer-Verglag. Fig. 276, p. 322.

Plate VIII. Scanning electronmicrograph of *Paramecium multimicronu-cleatum*. Dorsal view illustrating the coordinated beating of the cilia. Magnification: 600×. With permission of microscopists G. A. Antipa and E. B. Small. Previously published as Fig. 30, p. 52 in Corliss, J. O. 1979. *The Ciliated protozoa*, Second edition, Pergamon.

individual nerve cells can be approached by only a limited
number of techniques and are incompletely understood. Mutational probes of nerve function are difficult because of the consequences of mutation to the organism; mutational perturbations of
sufficient magnitude to be measured might be tolerable to the cell
but lethal to the animal. Biochemical analysis of functional
neurones is also restricted, because nerve cells are characteristically mixed with other kinds of cells and cannot be easily sorted
out for direct chemical analysis. The *in vitro* culture of neurocytes
is of course possible, but cells in tissue culture may not manifest
the important characteristics, or at least may require the development of a new phenomenology of cell behavior before analysis can
be undertaken.

The Paramecium model is based on a well-established
phenomenology that associates ciliary motion and membrane
potential with directly observable cellular movements. Naitoh,
Eckert, and their associates have provided a biophysical foundation for Jennings' avoiding reaction. When paramecia encounter
a barrier or sense a chemical incompatibility, their ciliary
membranes are depolarized, the cilia reverse their beat, and the
cells swim backward. After a short interval the membranes are
again hyperpolarized and recover their normal beat, and the cells
move forward again in a different direction (Fig. 16–2).

Kung's strategy was to collect mutants that perform this
stereotyped pattern in a modified way and to examine the genetic,
biophysical, and biochemical bases for the differences. One group
of mutants, for example, were unable to perform the avoiding
reaction under any conditions. They moved directly forward
through solutions that would elicit avoidance in normal cells and
thus could be selected for this property. They were designated as
pawn mutants and were found to be associated with one of three
different genetic loci. Some of the mutants are temperature sensitive and manifest the *pawn* phenotype only at high temperature.
Electrophysiological analysis shows that at least two of the classes
of mutants have membranes defective in their ability to transfer

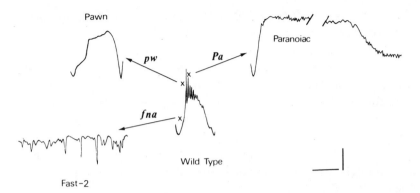

Fig. 16-2. Genetic modifications of active electrogenesis. Variant electrical activities can be viewed as impairments of the wild type activity by single point blockages resulting from genic mutations. The center of the figure shows one episode typical of the recording from wild type Paramecium, in which there is a long series of electrical responses to the appearance of 4 mM NaCl, 1 mM CaCl$_2$, and 1 mM Tris, pH 7.2, in the bath. In the case of the *fast-2* mutant, blockage near the beginning of the depolarizing activity by the *fna* mutation prevents depolarization completely. Blockage at the upstroke of the action potential by the *pw* mutation leads to the spikeless pattern of the *pawn* mutant. Blocking the downstroke of the action potential by the *Pa* mutation sustains the depolarization near the peak level for a much longer duration. Broken lines mark the estimated resting levels. Calibration: 10 Millivolts, 1 second. Reprinted by permission from Kung, C., S. Y. Chang, Y. Satow, J. van Houten, and H. Hansma. 1975. Genetic dissection of behavior in Paramecium. *Science,* **188,** 898–904. Copyright 1975 by the American Association for the Advancement of Science.

calcium in the depolarization reaction. These then are mutants of the calcium channel and provide a special means of investigating its function.

Another category of mutants shows "violent avoidances," and in the presence of sodium solutions may back continuously for 10–60 seconds, instead of transitorily as do the wild type. These overreacting mutants are called *paranoiac* and are characterized electrophysiologically as having prolonged membrane excitation.

Genetic analysis shows at least five genetic loci whose alteration can result in this phenotype; two of the loci are linked.

Still other mutants affect the rate of movement, responses to various ionic environments, and chemotactic responses. Such mutants may be selected by varying the environment and observing for atypical responses. A particularly useful discovery is that many *pawn* mutants are more tolerant of barium salts than are normal paramecia; these mutants may be selected as survivors in high barium solutions. Other mutants may be selected by their swimming rates through long tubes or by their passage through specially constructed mazes.

We cannot review this entire body of work, but recommend it as an example of the application of genetic dissection. It demonstrates the advantages of the large cell size (for electrophysiological studies), the microbial scale, the biochemical potentialities, and the mutational capacity of ciliates in the study of fundamental cellular function.

B. Nuclear Division and Cytokinesis

A second programmatic genetic dissection in ciliates has been begun by Frankel and his colleagues at the University of Iowa. This effort is concerned specifically with cell division in *T. thermophila* and more broadly with mechanisms of cellular morphogenesis. In the initial screening a total of 14 independent nitrosoguanidine induced temperature sensitive cell division mutants were identified. Genetic analysis distributed these among six complementation groups. Because the mutants yield at high temperature abortive dividers called *monsters,* the complementation groups were initially identified by characteristic mutants as *mo1, mo2, mo3,* and so on; they are now referred to as *cell division arrest (cda)* mutants. Each of these complementation groups is associated with a different genetic locus and with certain characteristic defects. All the members of a complementation group are not identical however. Certainly more than six genetic loci are involved in cytokinesis, but they may need to be detected by somewhat different procedures.

The *cdaA* mutants (Fig. 16-3) at the restrictive temperature of 40 °C are unable to divide, but they grow and make abortive attempts to divide. Stomatogenesis occurs, but the construction of the body and the interruption of the ciliary rows do not. The cell

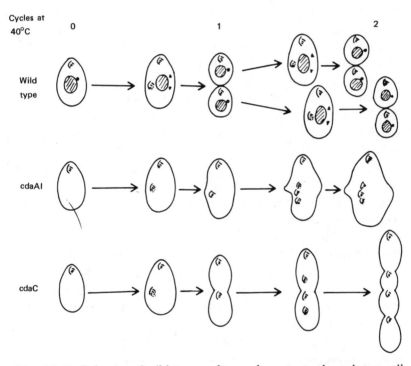

Fig. 16-3. Behavior of wild type and two *cda* mutants through two cell cycles at 40 °C. The nuclear cycles are shown only in the wild type. Reprinted, by permission, from Frankel, J., L. M. Jenkins, and L. E. DeBault. 1976. Causal relations among cell processes in *Tetrahymena pyriformis*: an analysis employing temperature sensitive mutants. *J. Cell Biol.* **71**, 242–260. Fig. 1, p. 245.

becomes progressively misshapen and accumulates a cluster of oral apparatuses. The macronucleus rarely divides after reaching the high temperature; even after a considerable period of time most of the blocked monsters contain only one macronucleus. In contrast, the micronuclear division cycle appears to be uncoupled so that it can continue more or less indefinitely, with increasing numbers of micronuclei dividing synchronously.

Mutants of the *cdaC* class are notable for the formation of tandem doublets. Again stomatogenesis occurs, but unlike in *cdaA* mutants a fission zone is formed and a constriction forms between the daughters. A second generation of oral primordia may appear and develop with functional gullets in semiisolated regions of a chain. Occasionally the chains break, but normal cytokinesis does not occur. The macronuclei in *cdaC* are more likely to divide than those in *cdaA* mutants, but few monsters come to have more than 2 or 3 macronuclei. The micronuclei, however, continue to divide, synchronously, for several cycles until many cells contain as many as 8, 16, 32, or more micronuclei.

Once the mutants have been individually obtained and characterized, they can be combined in various ways. When *cdaA/ cdaA*, *cdaC/cdaC* homozygotes are moved to the restrictive temperature, a small fraction of the cells quickly manifest the *cdaC* phenotype. The remainder of the cells eventually show a characteristic *cdaA* blockage. This result is interpreted in terms of gene action at different times. The $cdaA^+$ product is required early, and its lack is manifested in an early division block, prior to the stage of constriction. The $cdaC^+$ product acts later, after the formation of the fission zone and the beginning of constriction. In an unsynchronized population in fast exponential growth, some doubly homozygous cells will have accumulated sufficient $cdaA^+$ product to complete the early cytokinetic event, but they will be blocked at the later stage. Most cells, however, will have insufficient $cdaA^+$ product and will be blocked at the earlier stage.

We need not consider all the combinations of mutant classes to perceive the logic of their analysis. Three major categories of events are involved: those affecting the macronucleus, those affecting the micronucleus, and those observed in the cortical cytoplasm. Macronuclear division is closely associated with cytokinesis; micronuclear division is not. But macronuclear replication, as opposed to macronuclear division *is* relatively dissociated from cell division; DNA may continue to replicate for as many as four or five cycles within the single monstrous cell and without macronuclear division.

Assorting the effects of genetic blocks on cortical morphogenesis is more difficult. In the case of *cdaA* and *cdaC*, the behavior of the double homozygote suggests that *cdaA* action occurs before *cdaC* action; but this conclusion is indicated *a priori* by an analysis of the mutants individually. The data do not demonstrate that *cdaA* and *cdaC* are part of a single morphogenetic sequence in which each subsequent event depends on a prior event. The investigators conclude that *cdaA* and *cdaC* may be mechanistically independent and that they are arranged in sequence only by reference to a common temporal program.

Again, our object is not a critical evaluation or exhaustive summary, but an illustration of opportunities for analysis.

C. Phagocytosis

A third effort at genetic dissection is a cooperative venture of Eduardo Orias' group at the University of California, Santa Barbara, and Leif Rasmussen of the Carlsberg Laboratories in Copenhagen. The cellular function at issue is phagocytosis, or, in the context of the ciliates, the formation of food vacuoles. Phagocytosis in ciliates is limited to a special region at the base of the gullet. As in the case of the electrophysiological properties of the cell membrane, genetic probes of phagocytic capacity in multicellular organisms are not likely to be productive, mainly because significant membrane incapacities are likely to be fatal to the organism. With microbes it is possible to examine conditional membrane defects; even if the membrane defects are lethal under restrictive conditions, the mutants may be bred and analyzed. Moreover, in the case of Tetrahymena, it happens that the mutants may be maintained even under restrictive conditions if nutritional supplements are provided.

The selection of mutants begins with nitrosoguanidine mutagenesis, which is followed by a cross (to establish micronuclear mutations in the macronucleus where they may be expressed). The initially heterozygous mutants are allowed to grow (to allow allelic assortment and the expression of recessive mutants). Then the population is enriched for the desired phenotype by growing it alternately at 30 and 37 °C in the presence of tantalum. Wolfe has shown that feeding cells take up tantalum and can be physically separated from nonfeeding cells in a discontinuous Ficoll gradient. Hence the feeding cells at 30 °C are retained and placed at 37 °C, where the nonfeeding cells are retained. This process of enrichment is repeated several times. The final selection follows the addition of India ink to a 37 °C culture, and cells that do not form visible vacuoles are removed for study.

Most of the detailed work thus far has centered on one mutant designated as NP1. Gullet areas formed at low temperatures in this mutant continue to function at higher temperatures, but new oral apparatuses developed at higher temperatures, that is, the oral apparatuses of the posterior daughters, are nonfunctional (Fig. 16–4).

The defective cells under restrictive conditions are able to undergo a few fissions, but soon stop growing in ordinary media. They may grow indefinitely, however, if supplements of folinic acid, Fe, and Cu are provided. These supplements do not result in the formation of functional gullets, but permit sufficient nutrient uptake through the surface membranes (pinocytosis) to sustain growth.

These results, of course, provide only a beginning of a genetic analysis of phagocytosis, but they show that the techniques for

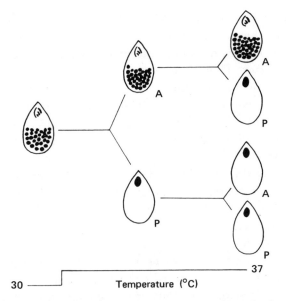

Fig. 16–4. The transmission of the capacity to form food vacuoles in mutant *NP1* of *Tetrahymena thermophila* after a temperature shift-up. The anterior and posterior daughters are indicated by *A* and *P*, respectively. The black circles represent the food vacuoles formed in the presence of India ink; their absence indicates a failure of food vacuoles to form. The filled oral region represents a nonfunctional oral apparatus. Reprinted with permission from Orias, E. and N. A. Pollock. 1975. Heat sensitive development of the oral organelle in a Tetrahymena mutant. *Exp. Cell Res.*, **90**, 345–356. Fig. 4, p. 350. Copyright by Academic Press Inc. (London) Ltd.

selecting mutations, as well as the means for their analysis, are at hand.

D. Exocytosis

A final example of genetic dissection again comes from *Paramecium tetraurelia* and is concerned with the development and extrusion of the trichocyst. Most ciliates contain extrusosomes of some kind, though their significance is not well understood. The trichocysts of Paramecium are located in regular arrays (Chapter 2) in the cortex. They do not arise *in situ,* however, but originate in cytoplasmic vesicles that grow, mature, and migrate to their characteristic cortical positions. The trichocysts are composed predominantly of a crystalline matrix, containing a protein with monomers of 17,000 daltons, and have a form something like a carrot, with the broad end attached to the cell membrane. When appropriately stimulated the trichocysts are discharged to the outside of the cell, taking the form of long threads with periodic striations and with a pointed tip at the end. A massive discharge can be triggered by picric acid or other agents to produce a marked halo effect around the cell, as well as to kill it.

Trichocyst mutants have been sought by isolating exautogamous clones from mutagenized cultures and exposing parts of the clones to picric acid. Numerous "nondischarge" mutants are easily collected and analyzed. Pollack, for example, describes 14 independently derived mutations that are distinguishable as 6 different phenotypes. One of the mutants forms no mature trichocysts though it produces pretrichocyst vesicles. The others develop trichocysts, usually of abnormal shape (*"football," "stubby," "pointless," "screwy-cigar"*), and these fail to attach to the pellicle, at least in normal numbers. Most of the mutations have secondary effects on other cell characteristics. Genetic analysis shows that at least 9 different loci are represented by the 14 mutations. Mutants at all except one locus, that associated with *screwy-cigar,* are also temperature sensitive, that is, unable to grow at higher temperatures.

A fine structure analysis of some of these mutants by Beisson and her colleagues provides further information concerning the mutant effects. Freeze-fracture studies show a characteristic pattern of membrane particles on the pellicle at the site of insertion of the trichocyst. A central rosette of particles 75 nanometers in diameter is surrounded by a double outer ring of particles 300 nanometers in diameter. A nondischarge mutant, *tam 8,* in which no trichocyst attachment occurs, was found to have no central rosette. More impressively, a conditional mutant *nd9* had normal rosettes at 18 °C, the permissive temperature for discharge, but had no rosettes or rosettes composed of only a few granules at 27 °C, the restrictive temperature. Finally, mutant *tl,* which produces no trichocysts, also produces cortical patterns with no rosettes. A rosette is apparently formed only when trichocyst insertion occurs. Failure of attachment may reflect the absence of a trichocyst (*tl*), the constitutional abnormality of the trichocyst (*tam 8*), or the conditional abnormality of some aspect of the interaction between the trichocyst and its attachment site (*nd 9*).

Another category of mutational defects of trichocysts has been explored by Aufderheide. As is noted earlier, the trichocysts do not develop in position, but in the solated cytoplasm. They are transported indiscriminately by the cyclotic motion of the endoplasm, but they also manifest jerky or "saltatory" movements of their own that aid in their eventual localization in the cortex. They can also be observed to "wobble" in curious ways as they orient to their final attachment points. Aufderheide has shown that some mutants are defective in their saltatory motions and others are unable to carry out the terminal insertion specific maneuvers.

Aufderheide was also able to discover whether failure of a trichocyst to insert is due to an aberration in the trichocyst itself or to an imperfection in the docking site. This issue was examined by taking advantage again of the large size of Paramecium and its lack of compartments. Cytoplasm was withdrawn from normal cells by pipette and reinjected into mutant cells. Alternatively, cytoplasm from noninserting mutants, containing trichocysts, was injected into normal cells and the behavior of the mutant trichocysts was observed. In the case of mutants *tam-8, ndA,* and *ndB,* the introduction of trichocysts into normal cells did not result in their insertion. On the other hand, if normal trichocysts are introduced into the cytoplasm of these mutants, the normal trichocysts are inserted and are capable of being discharged. The defects are associated with the trichocysts and not with the membrane. In contrast, if trichocysts from the nondischarge mutant *nd9* are injected into normal cells, they can be inserted and discharged. Moreover, the incapacity of *nd9* can be remedied by injection of normal cytoplasm; not only the normal trichocysts, but also the preexisting mutant trichocysts, can be ejected. Some cytoplasmic component essential for discharge is conditionally lacking in *nd9*.

Clearly, the synthesis, morphogenesis, movement, placement, insertion, and discharge of this single organelle are complex processes involving the activities of many different genes. Clear also are the virtues of a system of analysis that makes possible a progressive description and resolution of the mechanisms required.

SUMMARY

The ciliates have been sufficiently domesticated to permit the beginning of genetic dissection of several basic eukaryotic mechanisms. Four examples of such studies are introduced, two from *P. tetraurelia* and two from *T. thermophila*. The most ambitious of these is a dissection of membrane function in Paramecium, in which electrophysiology, cellular behavior, and ciliary function are the combined focus of mutational analysis. A second study with Paramecium is concerned with the origin, development, ultrastructure, insertion, and discharge of the extrusible trichocyst; again several different genetic loci affecting these processes have been identified. The first tetrahymena project approaches cytokinesis through mutational analysis, and the second explores phagocytosis through the use of conditional mutations.

RECOMMENDED READING

Aufderheide, K. J. 1978. The effective sites of some mutations affecting exocytosis in *Paramecium tetraurelia*. *Mol. Gen. Genet.,* **165,** 199–205.

Beisson, J., M. Lefort-tran, M. Pouphile, and M. Rossignol. 1976. Genetic analysis of membrane differentiation in Paramecium. *J. Cell Biol.,* **69,** 126–143.

Bruns, P. J. and Y. M. Sanford. 1978. Mass isolation and fertility testing of temperature-sensitive mutants in Tetrahymena. *Proc. Natl. Acad. Sci. U.S.,* **75,** 3355–3358.

Bruns, P. J., T. B. Brussard, and A. B. Kavka. 1976. Isolation of homozygous mutants after induced self-fertilization in Tetrahymena. *Proc. Natl. Acad. Sci. U.S.,* **73,** 3243–3247.

Byrne, B. J. and B. C. Byrne. 1978. Behavior and the excitable membrane in paramecium. *CRC Crit. Rev. Microbiol.* September.

Frankel, J., L. M. Jenkins, and L. E. DeBault. 1976. Causal relations among cell processes in *Tetrahymena pyriformis:* an analysis employing temperature sensitive mutants. *J. Cell Biol.,* **71,** 242–260.

Frankel, J., Jenkins, L. M., Doerder, F. P., and Nelsen, E. M. 1976. Mutations affecting cell division in *Tetrahymena pyriformis*. I. Selection and genetic analysis. *Genetics,* **83,** 489–506.

Kung, C., S.-Y. Chang, Y. Sato, J. Van Houten, and H. Hansma. 1975. Genetic dissection of behavior in Paramecium. *Science,* **188,** 898–904.

Nelson, D. L. and C. Kung. 1978. Behavior of Paramecium: chemical, physiological and genetic studies. In *Taxis and Behavior* (Receptors and Recognition, Series B, Vol. 5) (G. L. Hazelbauer, Ed.), Chapman and Hall, pp. 77–100.

Orias, E. and Bruns, P. J. 1976. Induction and isolation of mutants in Tetrahymena. *Methods Cell Biol.,* **13,** 247–282.

Orias, E. and N. A. Pollock. 1975. Heat sensitive development of the oral organelle in a Tetrahymena mutant. *Exp. Cell Res.,* **90,** 345–356.

Pollack, S. 1974. Mutations affecting the trichocysts in *Paramecium aurelia*. I. Morphology and description of the mutants. *J. Protozool.,* **21,** 352–362.

Schein, S. J. 1976. Nonbehavioral selection for Pawns, mutants of *Paramecium aurelia* with decreased excitability. *Genetics,* **84,** 453–468.

Van Houten, J., S-Y Chang, and C. Kung. 1977. Genetic analyses of "Paranoiac" mutants of *Paramecium tetraurelia*. *Genetics,* **86,** 113–120.

Conclusion | 17

THIS BOOK ORIGINATED IN A SERIES OF LECTURES entitled "The Organism as Technology," in which the ciliated protozoa were used to illustrate a theme. The thesis of the series was that the "domesticated" organism is an essential, but sometimes invisible, technology underlying much of our recent understanding of biological mechanisms. The domestication of animals as beasts of burden and the domestication of plants and animals as sources of food transformed human culture. Similarly, the domestication of organisms as instruments of scientific analysis has transformed the scientific enterprise and made possible dramatic biological advances of the present century.

Perhaps the first organism domesticated for scientific purposes was the fruit fly, and it continues as an essential component in our repertoire of biological tools. Drosophila was the "technique" used, first by the Morgan School, to establish the physical basis of transmission genetics, and it continues to be a precision tool in evolutionary studies and in developmental analysis. One basic technique—*Drosophila melanogaster*—provides the essential methodologies, but it is supplemented by a swarm of related species useful for special purposes.

Another fundamental technology, developed first by Dodge and Lindegren, is the red bread mold. Neurospora permitted Beadle, Tatum, and their associates to explore the means whereby genes affect biological specificities and led eventually to the demonstration that genes are the ultimate physiological agents, achieving their effects through the specification of proteins. As in the case of Drosophila, Neurospora technology centers on one species, *Neurospora crassa*, but it is supplemented and cross-illuminated by work on related species. As another ascomycete, even brewer's yeast, *Saccharomyces cerevisiae*, might be considered a part of the Neurospora technological battery—though it now threatens to become the dominant member.

Without question the central technological advance leading to our current profound understanding of molecular biological

289

mechanisms was the domestication by Lederberg, Luria, and Delbruck of the coliform bacterium, *E. coli,* and its viruses. This conclusion is perhaps sufficiently obvious as to require no elaboration.

An appropriate question at this point might be, How many scientifically domesticated organisms are now available? Perhaps, without too much argument, most biologists would agree that in addition to the organisms mentioned above, maize and the mouse are fully domesticated biological tools. *Zea mays,* like yeast, has another significance as a domesticated organism and was in fact domesticated first as a food plant; its scientific domestication came much later. The mouse, *Mus musculus,* has perhaps been "domesticated" in a way ever since it first moved into a human domicile, but its scientific domestication has been achieved only in this century, at great cost, and to great profit.

One might argue that these five organismic technologies provide an adequate foundation for modern experimental biology. Or one might want to consider a few other candidates. Other higher plants have certainly been used to advantage in scientific studies, but none is so well perfected for analysis as is maize. Again, other mammals such as the rat and the rabbit have made major contributions to our understanding of life processes, but those may be considered auxiliary technologies that borrow credibility from the premier organism—the mouse. And in any case, from our evolutionary perspective all the mammals including man are relatively minor variations on an organismic theme.

Without doubt scientific advances have been made using other organisms as experimental objects. For example the contributions made with echinoderms and with amphibians in developmental studies cannot be neglected. However, echinoderms are in no sense domesticated, and the long road to the domestication of amphibians has only begun.

Perhaps I should make more explicit what I mean by "scientific domestication." Certainly the ability to propagate an organism indefinitely "in captivity" is an essential step toward providing a reliable supply of experimental material. It is also an important step toward controlling the biological variables in experimental manipulations. Much more is needed, however, if we are to manipulate in the most precise and subtle ways the sources of biological specificity. Full domestication requires the ability to manipulate genetic variables and, indeed, to induce and select mutations affecting the mechanisms being explored (Chapter 16). The maturity of an organism as a domesticated technology must be measured against its capacity for genetic dissection.

Genetic dissection is not, of course, the only successful route to biological mechanisms. It is perhaps the extreme refinement of the comparative approach, which is the most characteristically "biological" technique for the identification of significant variables. Before the twentieth century it was usually applied on a broad organismic basis, but genetic dissection now allows comparisons to be made on the effects of subtle intrinsic chemical substitutions in otherwise identical reaction systems. Genetic dissection does not, however, deny the value of other experimental strategies or eliminate the need for cruder comparisons.

If we conclude that modern experimental biology employs five major organismic technologies (or constellations of technologies), does it need any more? This question is perhaps similar to one concerning the necessity of additional food plants. We are surviving on the present resources, but life could always be better. The domestication of food plants has historically followed a progression from the gathering of wild fruits, through the development of "folk varieties" under partial human control, to the commercial development of genetically uniform plants for mass production. Many potential food plants progress only part of the way to full domestication. They may prove refractory at some stage, or unable to compete economically with alternative food sources, or unpalatable to many consumers. The usefulness of new forms cannot be measured until the process of domestication has been pursued somewhat further. Generally, however, successful new forms must offer either novel characteristics or competitive advantages.

The necessity for novelty makes unlikely the full domestication of new species closely related to the "big five." Organisms closely related to *Drosophila melanogaster* or *E. coli* will likely become satellites of their relatives. Fundamentally new technologies must emerge from distinctively different kinds of organisms. Among incipient major technologies now emerging we must mention *Chlamydomonas reinhardii* from the green algae, *Coenorhabditis elegans* from the nematodes, and perhaps Physarum and Dictyostelium from the slime molds.

In the preceding chapters I have set forth some of the evidences that the ciliated protozoa may also be approaching maturity as organismic technologies. Much of their use in the past has been as comparative foils, means of testing the generality of mechanisms of life processes discovered in other systems. This use will certainly continue. However, the ciliates have also provided the basis for several significant original observations. I do not here attempt a summary of ciliate contributions, but call attention to one central issue—that of the maintenance of biological specificity in clonal proliferation. The ciliates provide several examples in which different cellular characteristics may be specifically induced in cells of identical genotype and subsequently maintained indefinitely in clonal culture without being associated with conventional structural changes in the nuclei. These studies go beyond the demonstration of such phenomena to the discovery of different classes of mechanisms for the perpetuation of differences (Chapters 10–15), though without satisfactorily resolving the details of the mechanisms. I hope that the rapidly improving ciliate technology will soon provide a firmer mechanistic basis for these cases of cellular differentiation, and exportable understanding about fundamental eukaryotic processes.

Acknowledgements

18

PEDAGOGICAL SIMPLIFICATION, LIKE ALL GENERAL-
ization, is a hazardous business, and its successful accomplish-
ment requires profound understanding, extraordinarily good luck,
or well-informed and long suffering friends. The necessity for
familiarity with the phenomena treated explains in part, though it
may not excuse, the choice of much material from my own
experience, even when more apt illustrations were available
elsewhere. Of course I cannot rely on my own experience in many
areas and have had to venture into less familiar territory where
errors—from slips of the pen to major misunderstanding—are very
probable and can seriously flaw the exposition. What merits the
final result may have derive in large measure from its communal
nature. Residual errors certainly remain, but most of these are
secondary errors, mistakes made in the process of repairing
primary errors identified by the patient experts. Some of these
collaborators read carefully and criticized tactfully the whole
book, in critiques sometimes running longer than the chapters.
These include André Adoutte, Peter Bruns, Paul Doerder, Joseph
Frankel, Marlo Nelsen, and Tracy Sonneborn. Other experts
induced to give special attention to particular portions of the
manuscript or more global scrutiny of the whole were Lea Bleyman,
Ruth Dippell, Klaus Heckmann, Linda Maxson, Akio Miyake,
Dennis Nyberg, Carl Woese, and of course all the local crew in
experimental ciliatology at the University of Illinois at Urbana-
Champaign.

Author Index

Subject Index